Tropical Forestry

Series editor
Michael Köhl
Hamburg, Germany

Tropical forests are disappearing with dramatic speed. The causes of destruction of tropical forests are extremely complex and differ not only country by country, but also due to economic, political and social factors. The series Tropical Forestry publishes volumes on relevant disciplines which are essential for successful and sustainable forestry activities. It is addressed to forestry professionals, natural resource managers and all those who are interested in the restoration, conservation and management of tropical forestry resources in academia and in the corporate world.

More information about this series at http://www.springer.com/series/5439

Rozi Mohamed
Editor

Agarwood

Science Behind the Fragrance

Springer

Editor
Rozi Mohamed
Universiti Putra Malaysia
Department of Forest Management
UPM Serdang
Malaysia

ISSN 1614-9785
Tropical Forestry
ISBN 978-981-10-0832-0 ISBN 978-981-10-0833-7 (eBook)
DOI 10.1007/978-981-10-0833-7

Library of Congress Control Number: 2016942099

Printed on acid-free paper

This Springer imprint is published by Springer Nature
The registered company is Springer Science+Business Media Singapore Pte Ltd.

Preface

Agarwood is the contemporary name given to this fragrant natural product from trees of the Thymelaeaceae family, the most widespread being *Aquilaria*. This ancient product has provided a great sense of appreciation to humankind in many ways, from incenses used in rituals by different societies around the world to traditional medicines and fragrances. In modern days, agarwood still performs its long acclaimed services but with a renewed interest as it gains popularity in cosmetics and medicine manufacturing. Being heavily sourced from the wild, agarwood resources are quickly depleting due to habitat demolition and destructive harvesting techniques. In recent years, *Aquilaria* plantations have received increasing attention as a renewable source of agarwood for various downstream industries.

Agarwood fragrance is distinctive that it is often associated with myths and spiritual experiences. This book is dedicated to understanding the facts about agarwood formation, hence the title *Science Behind the Fragrance*. Despite its centuries-old existence, the biology of agarwood resources has been very much eluded. Domestication efforts of the tree provide promising new developments in agarwood induction and detection technologies. Scientists have identified critical components that affect agarwood productivity resulting from the tree host reaction to biotic and abiotic stresses. Advances in biotechnology and genomics of *Aquilaria* species have shed light on the process of its cell machinery in synthesizing crucial compounds in agarwood, many of which are responsible for the unique fragrance. New tree resources have been explored, and new pharmaceutical properties have emerged. All these are covered in the chapters, providing researchers and the public with a sourcebook of current knowledge on agarwood.

Most of the chapter authors participated in the 1st International Scientific Symposium on Agarwood (ISSA), *Agarwood in the New Era*, convened in 2013 at the Universiti Putra Malaysia (UPM), Serdang campus. Organized by the Faculty of Forestry, UPM, with collaboration from the Forest Research Institute Malaysia (FRIM), the Asia Pacific Association of Forestry Research Institutions (APAFRI), the Malacca State Forestry Department, and the Malaysian Timber Industry Board (MTIB), together with compassionate sponsors, ISSA2013 provided a platform for the genesis of this book.

I especially thank all chapter authors for their valuable contributions. This book was designed to nurture information exchange and to inspire agarwood scientists. I hope this publication will serve as a basis for future research with agarwood tree species and other aromatic tropical tree genera.

UPM Serdang, Malaysia
January 2016

Rozi Mohamed

Contents

Contributors

Mengyu Dong Department of Wood Anatomy and Utilization, Research Institute of Wood Industry, Chinese Academy of Forestry, Beijing, P.R. China

Zhi-Hui Gao Institute of Medicinal Plant Development, Chinese Academy of Medical Science & Peking Union Medicinal College, Beijing, P.R. China

Hideaki Hara Molecular Pharmacology, Department of Biofunctional Evaluation, Gifu Pharmaceutical University, Gifu, Japan

D.S. Hettiarachchi Wescorp Group of Companies, Canning Vale, WA, Australia

Asep Hidayat Forest Plantation Fiber Technology Research Institute, FORDA, Ministry of Environment and Forestry, Kuok, Riau Province, Indonesia

Xiaomei Jiang Department of Wood Anatomy and Utilization, Research Institute of Wood Industry, Chinese Academy of Forestry, Beijing, P.R. China

Lichao Jiao Department of Wood Anatomy and Utilization, Research Institute of Wood Industry, Chinese Academy of Forestry, Beijing, P.R. China

Mamoru Kakino Molecular Pharmacology, Department of Biofunctional Evaluation, Gifu Pharmaceutical University, Gifu, Japan

Lina Karlinasari Department of Forest Products, Faculty of Forestry, Bogor Agricultural University (IPB), Bogor, Indonesia

Shiou Yih Lee Department of Forest Management, Faculty of Forestry, Universiti Putra Malaysia, UPM, Serdang, Selangor, Malaysia

Rozi Mohamed Department of Forest Management, Faculty of Forestry, Universiti Putra Malaysia, UPM, Serdang, Selangor, Malaysia

Che Mohd Aizal Bio Aromatic Research Center of Excellence, Universiti Malaysia Pahang, Gambang, Pahang, Malaysia

Dodi Nandika Department of Forest Products, Faculty of Forestry, Bogor Agricultural University (IPB), Bogor, Indonesia

Saiema Rasool Department of Forest Management, Faculty of Forestry, Universiti Putra Malaysia, UPM, Serdang, Selangor, Malaysia

Erdy Santoso Centre for Conservation and Rehabilitation, Forest Research and Development Centre (FRDC), Forestry Research and Development Agency (FORDA), Ministry of Environment and Forestry, Bogor, Indonesia

S.M.C.U.P. Subasinghe Department of Forestry and Environmental Science, University of Sri Jayewardenepura, Nugegoda, CO, Sri Lanka

Saiful Nizam Tajuddin Bio Aromatic Research Center of Excellence, Universiti Malaysia Pahang, Gambang, Pahang, Malaysia

Maman Turjaman Centre for Conservation and Rehabilitation, Forest Research and Development Centre (FRDC), Forestry Research and Development Agency (FORDA), Ministry of Environment and Forestry, Bogor, Indonesia

Jian-He Wei Department of Culture Centre, Institute of Medicinal Plant Development, Chinese Academy of Medical Science & Peking Union Medicinal College, Beijing, P.R. China

Hainan Branch, Institute of Medicinal Plant Development, Chinese Academy of Medical Science, Haikou, P.R. China

Hainan Provincial Key Laboratory of Resources Conservation and Development of Southern Medicine, Wanning, P.R. China

Yafang Yin Department of Wood Anatomy and Utilization, Research Institute of Wood Industry, Chinese Academy of Forestry, Beijing, P.R. China

Mashitah Mohd Yusoff Faculty of Industrial Sciences and Technology, Universiti Malaysia Pahang, Gambang, Pahang, Malaysia

Shujuan Zhang Beijing Tian-yi-li-hua Institute of Agarwood, Beijing, P.R. China

Chapter 1
The Origin and Domestication of *Aquilaria*, an Important Agarwood-Producing Genus

Shiou Yih Lee and Rozi Mohamed

Abstract The *Aquilaria* (Thymelaeaceae) tree is a well-known important agarwood-producing genus, which is endemic to the Indomalesia region. The genus is currently protected under CITES regulation and the IUCN Red List due to its heavy declination in the natural population in various sourcing countries. Derived from its precious non-wood fragrant products, the genus was given different names throughout the history until it was finalized in 1783. To date, there are 21 recognized *Aquilaria* species recorded, of which 13 are reportedly fragrant resin producers, and the status of the remaining eight *Aquilaria* species is yet to be investigated. *Aquilaria* is heavily exploited in the wild due to the destructive agarwood harvesting technique that requires hacking of the wood parts to induce agarwood production. Various conservation efforts have been carried out to avoid further destruction toward its gene pool. This includes introducing the species for cultivation and planting the trees in large plantations or home gardens, which further provide a sustainable agarwood production in the industry and indirectly contribute to the local economy. At present, an accurate classification of *Aquilaria* species is yet to be achieved; misidentification happens frequently, either genuinely because of lack of information and training or intentionally for business gains. In conclusion, a proper taxonomy and classification system are essential for conserving *Aquilaria* species genetic diversity and for identifying species origin of agarwood products aimed at international trade control.

1.1 Introduction

The *Aquilaria* genus is well known for its fragrant non-wood product, the agarwood. Highly demanded in several countries, agarwood is further processed into perfumes, incenses, and ornamental displays and used as a raw material in traditional and modern medicines. Historically, human's encounter with agarwood was first

S.Y. Lee • R. Mohamed (✉)
Department of Forest Management, Faculty of Forestry, Universiti Putra Malaysia, UPM, Serdang, Selangor, 43400, Malaysia
e-mail: rozimohd@upm.edu.my

R. Mohamed (ed.), *Agarwood*, Tropical Forestry,
DOI 10.1007/978-981-10-0833-7_1

recorded in ancient literatures and religious scriptures. The Sanskrit language poet, Kâlidâsa (c. 353–c. 420), once wrote: *Beautiful ladies, preparing themselves for the feast of pleasures, cleanse themselves with the yellow powder of sandal, clear and pure, freshen their breast with pleasant aromas, and suspend their dark hair in the smoke of burning aloes*. The word "aloes" has the same meaning as agarwood. It was also found occurring in the biblical text (Duke 2007). In ancient days, the Egyptians used agarwood to embalm honored dead bodies, and in several Asian countries, agarwood products were introduced along with Buddhism from India. During those days, the most familiar *Aquilaria* species producing agarwood were *Aquilaria agallocha* (synonym to *Aquilaria malaccensis*) from India and nearby countries and *Ophispermum sinense* (synonym to *Aquilaria sinensis*) from China (Don 1832). The former was widely applied in fragrance production and religious practices while the latter in Chinese medicines.

From a total of 21 accepted species names for *Aquilaria* at present (The Plant List 2013), about thirteen are reported as fragrant resin producer: *A. baillonii*, *A. beccariana*, *A. crassna*, *A. filaria*, *A. hirta*, *A. khasiana*, *A. malaccensis*, *A. microcarpa*, *A. rostrata*, *A. rugosa*, *A. sinensis*, *A. subintegra* and *A. yunnanensis* (Hou 1960; Ng et al. 1997; Compton and Zich 2002; Kiet et al. 2005; Yang Y 2015, personal communication). As for the remaining eight *Aquilaria* species, their competencies at producing agarwood need to be further investigated.

1.2 A Brief History of the Genus

The genus name, *Aquilaria*, was originally derived from its non-wood fragrant product, presently known as agarwood. Agarwood has many names that have been passed down over many generations. When the product is introduced into various societies, each mimicked the pronunciation of the original name using her own language; thus, more names were created. From recorded history, the earliest name given was *ahalim* in Hebrew and *ahila* from the Scripture of the East (Ridley 1901), followed by *agalukhi* in Arabic. It was also described as *agallochee*, a Greek synonym to a Hindi word for incense wood *aod-i-kimaree* from India and Arabia. In the Malayan region, it has been known as *agila*, which possibly descended from the Sanskrit *agara* (in Hindi *aggur*). In more recent days, the Portuguese gave several names: *pao-d'agila*, *pao-d'aguila*, *pao-d'aquila*, *bois d'aigle*, *eagle-wood*, and *agel-hout*. The genus established its final name, *Aquilaria*, in 1783, given by the botanist Jean-Baptiste Lamarck after replacing its synonym, *Agallochum* of Dioscorides.

The first scientific record of agarwood usage was likely that of Avicenna, an Arab physician (980–1037), who described several types of *agallochum* (Society for the Diffusion of Useful Knowledge 1838). Among the different types of *agallochum*, he recorded two names, *xylaloes* and *agalugen*. *Xylaloes* is a Greek form of an Arabic word *alud*, literary described as "the wood." Further, it was modified into aloe wood/aloes-wood and also Lignum aloes. As for *agalugen*, it was also called *aghaloojee*, which was then defined as *agallochee* or *agallochum*. The word *agallochum* was simplified to *agalloch*, referring to the fragrant wood produced from *A. agallocha* in India.

The first formal account to the tree itself was by Garcia de Orta (1501–1568), a Portuguese Renaissance physician and naturalist, who practiced in Goa, India, and was a pioneer of tropical medicine (Ridley 1901). He visited Malacca in Peninsular Malaysia roughly in 1534 and named the fragrant wood *garo*. He recorded that the wood was brought by the Chinese from Malacca and Sumatra; thus, he referred it as *Garo de Malacca*. Garcia successfully collected the twigs and leaves from trees growing in Malacca but failed to get the fruits or flowers as there was difficulty to access the forest. Georg Eberhard Rumphius (1627–1702), a German-born botanist who was studying on the specimen brought back from Malacca, distinguished two types of *agallochum*: the *calambac* (*Agallochum primarium*) and the *garo* (*Agallochum secundarium*) (Society for the Diffusion of Useful Knowledge 1838). The *calambac* or *calembouc* (French) had other names such as *kỳ nam* (Vietnam) and *kyara* (Japanese) (Li 1998). The first collection of *calambac* was native to Eastern Cochin China and Siam, collected by Loureiro from the tree called *Aloexylum agallochum* Lour., while the second collection was native to Cochin China and Laos, also collected by Loureiro from the tree called *Ophispermum sinense* Lour. (known as *A. sinensis* at present) (Ridley 1901). William Roxburgh (1751–1815), a Scottish surgeon and botanist, who is known as the Father of Indian Botany, described that the real *calambac* comes from *A. agallocha Roxb.*, which was exported to China from the eastern frontier instead of Cochin China. *Calambac* from both origins had equal demands and were growing at similar latitudes, yet no one could conclude that they derived from the same species because botanical description was incomplete at that time (Society for the Diffusion of Useful Knowledge 1838).

Garo was later known as *garos*, which was recorded as an article of export from Malacca and the kingdom of Siam (Thailand at present). It was given the name *garu* (in Malay, later updated to *gaharu*), derived from the Sanskrit *aquaru* but only referring to the fragrant wood. The *garu* tree is given a different name and is known as *karas*, *tuikaras*, *tengkaras*, *engkaras*, and *kakaras* by the Malays (Ridley 1901). Pierre Sonnerat (1748–1814), a French naturalist and explorer, successfully obtained the specimens of the tree during his second voyage to India, based on the figures and description done by Jean-Baptiste Lamarck (1744–1829) for the *bois d'aloes*, *Agallochum officinarum*. Lamarck concluded that the collected specimen greatly resembles the *A. secundarium* from Rumphius; thus, it was renamed as *A. malaccensis* (Royle 1839). Upon confirmation, Francis Hamilton (1762–1829, also known as Francis Buchanan) concluded that *A. malaccensis* and *A. agallocha* are both of the same tree in nature but prefers the name *A. officinarum* as the official name for the plant (Hamilton 1836). Today, the name *A. malaccensis* retains as a type specimen for the genus *Aquilaria* in taxonomy identification. Since the official genus *Aquilaria* was agreed upon in 1783, it replaced several synonym genera including *Agallochum* Lam. (1783), *Aloexylum* Lour. (1790), *Aquilariella* Tiegh. (1893), *Decaisnella* Kuntze (1891), *Gyrinopsis* Decne. (1843), and *Ophispermum* Lour. (1790) (Tropicos 2016).

1.3 Generic Status and Relationships

Aquilaria is a member of the Thymelaeaceae (Malvales) family and belongs to the subfamily Thymelaeoideae (previously Aquilariodeae). There was a minor controversy in the classification of the subfamilies in Thymelaeaceae. Until today, a stable circumscription in the taxon of Thymelaeaceae is yet to be achieved. Conventional classification of the Thymelaeaceae is always related to the identification through morphological and reproductive characteristics of the plant itself. Earlier in 1836, before an international taxonomy system was established, *Aquilaria* was once under the order Aquilariaceae, Alliance Daphnales. However in 1880, the Bentham and Hooker system removed the genus on reasons that there was no recollection exercise since previous identification (Watt 2014). Years later, it was re-added, together with *Gyrinopsis* and *Gyrinops* into the family Thymelaeaceae, under subfamily Aquilarioideae (including Phalerioideae, Thymelaeoideae, and Drapetoideae) and tribe Aquilarieae (Gilg 1894). Thymelaeaceae was under the order of Thymelaeales. Later in 1967, Hutchinson proposed to include Aquilariaceae under Thymelaeales. However in 1968, Cronquist proposed to embed Thymelaeaceae under the order Myrtales, then further suggested by Thorne to place under Euphorbiales, instead of Myrtales. The debate between Myrtales and Euphorbiales was ongoing from 1968 to 1993, with a small different opinion whereby Cronquist's proposal was accepted from 1988 to 1992, in which Thymelaeaceae was placed as the sole family under the order Thymelaeales (Cronquist 1988). In 1993, Heywood placed Thymelaeaceae under the order Myrtales after removing it from the previous order Euphorbiales, which was proposed by Thorne in 1992. In 1998, the Angiosperm Phylogeny Group included the family Thymelaeaceae in the Malvales order, disregarding indication toward families adjacent to it (Angiosperm Phylogeny Group 1998). Chronology of the events is shown in Table 1.1.

Subfamilies within the Thymelaeaceae were not established as well throughout the centuries. As such in 1894, Gilg proposed four subfamilies under Thymelaeaceae: Aquilarioideae, Phalerioideae, Thymelaeoideae, and Drapetoideae. However in 1921, he reviewed the list and added three new subfamilies: Microsemmatoideae, Octolepidoideae, and Synandrodaphnoideae. Domke (1934) scaled down the list by retaining two names and adding two new ones: Aquilarioideae, Gilgiodaphnoideae, Gonystyloideae, and Thymelaeoideae. The latest update in the subfamilies within the Thymelaeaceae was by Herber in 2002 and 2003, concluding only two big subfamilies: Octolepidoideae and Thymelaeoideae, with Aquilarieae placed under Thymelaeoideae. At present, *Aquilaria* and its closely related genus *Gyrinops* are under the order Malvales, family Thymelaeaceae, subfamily Thymelaeoideae, and tribe Aquilarieae.

The four-subfamily classification, proposed by Gilg (1894), was mostly supported by molecular phylogeny as revealed from the sequences of the chloroplast DNA, the *rbc*L gene and the *trn*L-*trn*F intergenic spacer region, of forty-one samples under the Thymelaeaceae (Van der Bank et al. 2002). Unfortunately, the phylogenetic relationship was not supported at level of tribe. The study was further investigated by

Table 1.1 Chronology of events following changes in taxonomy affinities of the genus *Aquilaria* (Thymelaeaceae)

Taxonomist	Year proposed		Changes made
Lidney	1836	Alliance Order Genera	**Daphnales** **Aquilariaceae**; Elaeagnaceae; Hernandiaceae; Thymelaeaceae; ***Aquilaria***; *Gyrinops*; *Ophiospermum*
Bentham and Hooker	1880	Genera	***Aquilaria*** was removed
Gilg	1894	Subfamily Tribe Genera	**Aquilarioideae**; Phalerioideae; Thymelaeoideae; Drapetoideae **Aquilarieae** ***Aquilaria***; *Gyrinopsis*; *Gyrinops*
Gilg	1921	Subfamily	**Aquilarioideae**; Phalerioideae; Thymelaeoideae; Drapetoideae; Microsemmatoideae; Octolepidoideae; Synandrodaphnoideae
Domke	1934	Subfamily General Tribe Genera	**Aquilarioideae**; Gilgiodaphnoideae; Gonystyloideae; Thymelaeoideae **Aquilariidae** ***Aquilaria***; *Gyrinops*
Hutchinson	1959	Order	**Thymelaeales**
Hutchinson	1967	Order Family Genera	**Thymelaeales** **Aquilariaceae**; Thymelaeaceae ***Aquilaria***; *Deltaria*; *Gyrinops*; *Lethedon*; *Octolepis*; *Solmsia*
Cronquist	1968	Order	**Myrtales**
Thorne	1968	Order	**Euphorbiales**
Cronquist	1988	Order	**Thymelaeales**
Thorne	1992	Order	**Euphorbiales**
Heywood	1993	Order	**Myrtales**
The Angiosperm Phylogeny Group	1998	Order	**Malvales**
Herber	2002	Subfamily Tribe Genera	Octolepidoideae; **Thymelaeoideae** **Aquilarieae**; Daphneae; Synandrodaphneae ***Aquilaria***; *Gyrinops*

Rautenbach (2008) who performed molecular phylogenetic analysis using 143 specimens from the Thymelaeaceae members, which was three times greater in sample size compared to the previous study. The two regions from the chloroplast DNA were analyzed in addition to the nuclear ribosomal DNA internal transcribed spacer (ITS). The results were in support of the classification proposed by Herber in 2002. Considering that molecular phylogeny approach can provide remarkable results that complement conventional taxonomic classification in Thymelaeaceae, it has been applied for identifying agarwood-producing species from the genus *Aquilaria*. Ito and Honda (2005) sequenced the ITS1 and *psb*C-*trn*S regions in their study, while Eurlings and Gravendeel (2005) sequenced the *trn*L-*trn*F intergenic spacer region.

Both studies used authentic samples from herbarium specimens and concluded that molecular-based approach is possible for identifying *Aquilaria* species. Eurlings and Gravendeel (2005) provided a wider scope when they included *Gyrinops* specimens and concluded that *Aquilaria* and *Gyrinops* are paraphyletic, indicating that the two genera had shared the last common ancestor. Although molecular-based study seems promising in assisting identification at genus and species levels, further studies are needed for conservation and trade control purposes.

1.4 Distribution of the Species

Aquilaria is widely distributed in the Indomalesia region. The most dominant species, which has its population over several countries, is *A. malaccensis*. The accepted species according to The Plant List, their distributions based on previous records, and their conservation status as classified by IUCN, are compiled in Table 1.2.

To help illustrate the distribution, an imaginary horizontal line parallel to the equator is drawn going from across the Sumatra Island to Borneo Island, and a vertical line is drawn from the east of Taiwan going through the west of the Philippines, separating Borneo from Sulawesi and west of Sumba Island (Fig. 1.1). For the benefit of this discussion, these crossing lines divide the Indomalesia region into four sections and reflect the distribution of the related species in a congruent manner. Starting with the northwest end, this first region is widely populated by *A. crassna*, *A. malaccensis*, and *A. sinensis*. The distribution of *A. crassna* has been reported in Cambodia, south of Laos, north of Thailand, and Cochin China of Vietnam; *A. malaccensis* in Bangladesh, Bhutan, Assam of northeast India, Sumatra and Kalimantan of Indonesia, Iran, Malaysia, Myanmar, south of the Philippines, Singapore, and south of Thailand; and *A. sinensis* meanwhile endemic to China, confined mainly to the south, Hainan Island, Hong Kong, and Taiwan. Besides that, records have shown that *A. baillonii* is endemic to Cambodia; *A. banaensis* to Bana of Vietnam; *A. beccariana* to East Malaysia, Brunei, and Kalimantan of Indonesia; *A. hirta* to south of Thailand and northeast and south of Peninsular Malaysia including Singapore; *A. khasiana* to Khasi, Meghalaya, of northeast of India; *A. rostrata* to Peninsular Malaysia; *A. rugosa* to Kontum of Vietnam and north of Thailand; *A. subintegra* to south of Thailand; and *A. yunnanensis* to Yunnan in China.

Interestingly, the Philippines, situated in the northeast, is the only country within that region having six endemic species: *A. brachyantha* in Cagayan, *A. decemcostata* in Laguna, and *A. parvifolia* in Camarines, all three species being concentrated in the Luzon Island, while *A. apiculata* in Bukidnon and *A. citrinicarpa* and *A. urdanetensis* in Mount Urdaneta, all in Mindanao, the south island. In addition, *A. cumingiana* was also recorded in Mindanao, besides the Maluku Island of neighboring Indonesia. Following the horizontal line, *A. microcarpa* was recorded in Johor, the most southern state in Peninsular Malaysia and in Singapore. It was also found on the Borneo Island, consisting of East Malaysia, Brunei, and Kalimantan of Indonesia.

Table 1.2 Accepted names, distribution, and conservation status of *Aquilaria* species

No	Species names[a]	Basionyms and synonyms	Year first reported	Distribution	Conservation status (IUCN)
1	*Aquilaria apiculata* Merr.	–	1922	Philippines	–
2	*Aquilaria baillonii* Pierre ex Lecomte	–	1915	Cambodia	–
3	*Aquilaria banaensis* P.H.Hô	*Aquilaria banaensae*	1986	Vietnam	1998: Vulnerable D2
4	*Aquilaria beccariana* Tiegh.	*Aquilaria cumingiana var. parvifolia* *Aquilaria grandifolia* *Gyrinops brachyantha* *Gyrinopsis grandifolia*	1893	Brunei Indonesia Malaysia	1998: Vulnerable A1d
5	*Aquilaria brachyantha* (Merr.) Hallier f.	*Gyrinopsis brachyantha*	1922	Philippines	–
6	*Aquilaria citrinicarpa* (Elmer) Hallier f.	*Gyrinopsis citrinicarpa*	1922	Philippines	–
7	*Aquilaria crassna* Pierre ex Lecomte	–	1914	Cambodia Laos Thailand Vietnam	1998: Critically Endangered A1cd
8	*Aquilaria cumingiana* (Decne.) Ridl.	*Aquilaria pubescens* *Decaisnella cumingiana* *Gyrinopsis cumingiana* *Gyrinopsis cumingiana var. pubescens* *Gyrinopsis decemcostata* *Gyrinopsis pubifolia*	1922	Indonesia Philippines	1998: Vulnerable A1d
9	*Aquilaria decemcostata* Hallier f.	–	1922	Philippines	–
10	*Aquilaria filaria* (Oken) Merr.	*Aquilaria cuminate* *Aquilaria tomentosa* *Gyrinopsis acuminate* *Pittosporum filarium*	1950	Indonesia Philippines	–
11	*Aquilaria hirta* Ridl.	*Aquilaria moszkowski*	1901	Indonesia Malaysia Singapore	1998: Vulnerable A1d

(continued)

Table 1.2 (continued)

No	Species names[a]	Basionyms and synonyms	Year first reported	Distribution	Conservation status (IUCN)
12	*Aquilaria khasiana* Hallier f.	–	1922	India	–
13	*Aquilaria malaccensis* Lam.	*Agallochum malaccense* *Aloexylum agallochum* *Aquilaria agallocha* *Aquilaria agallochum* *Aquilaria ovate* *Aquilaria moluccensis* *Aquilaria secundaria* *Aquilariella malaccense* *Aquilariella malaccensis*	1783	Bangladesh Bhutan India Indonesia Iran Malaysia Myanmar Philippines Singapore Thailand	1998: Vulnerable A1cd
14	*Aquilaria microcarpa* Baill.	*Aquilaria borneensis* *Aquilariella borneensis* *Aquilariella microcarpa*	1875	Indonesia Malaysia Singapore	1998: Vulnerable A1d
15	*Aquilaria parvifolia* (Quisumb.) Ding Hou	*Gyrinopsis parvifolia*	1960	Philippines	–
16	*Aquilaria rostrata* Ridl.	–	1924	Malaysia	1997: Vulnerable 1998: Data Deficient 2012: Critically Endangered B1ab(v)
17	*Aquilaria rugosa* K.Le-Cong & Kessler	–	2005	Thailand Vietnam	–
18	*Aquilaria sinensis* (Lour.) Spreng.	*Agallochum grandiflorum* *Agallochum sinense* *Aquilaria chinensis* *Aquilaria grandiflora* *Aquilaria ophispermum* *Ophispermum sinense*	1825	China	1997: Vulnerable 1998: Vulnerable B1+2cde

Table 1.2 (continued)

No	Species names[a]	Basionyms and synonyms	Year first reported	Distribution	Conservation status (IUCN)
19	*Aquilaria subintegra* Ding Hou	–	1964	Thailand	–
20	*Aquilaria urdanetensis* (Elmer) Hallier f.	*Gyrinopsis urdanetense* *Gyrinopsis urdanetensis*	1922	Philippines	–
21	*Aquilaria yunnanensis* S.C. Huang	–	1985	China	–

[a]According to Version 1.1 of The Plant List (2013), http://www.theplantlist.org, assessed on 26 January 2016
– not available

Fig. 1.1 A map of the Indomalesia region. The imaginary lines are drawn to illustrate the distribution of *Aquilaria* species into four parts of the region

In the southwest end, endemic species has not been reported. The common *Aquilaria* species are again *A. malaccensis* in Sumatra and Kalimantan of Indonesia, *A. beccariana* in Kalimantan, and *A. microcarpa* in Singapore and Kalimantan. Meanwhile, the southeast region consists of mainly West Papua of Indonesia and Papua New Guinea, where *A. filaria* dominates.

1.5 Conservation Status

In modern days, beginning from the twentieth century, *Aquilaria* receives great attention from societies around the world because of agarwood's economic value. Demand for agarwood increased when the market flows extensively into the Arab society for perfumery, Indian society for religious application, and Chinese society for incense and medicinal purposes, triggering the needs to search for more of these agarwood-producing trees. The production of agarwood was not consistent, and extensive harvesting of the trees threatened the reproduction cycle in its natural environment. The undesirable phenomenon draws the attention of the Convention of International Trade in Endangered Species of Wild Fauna and Flora (CITES) to control the trade of agarwood by restricting the quota of goods exported from each country. When *Aquilaria* tree was first exploited, much of the demand in the agarwood market was sourced by *A. malaccensis*, threatening its sustainability in the wild. Consequently, the species became the first *Aquilaria* listed in the Appendix II of CITES, bringing its status to "potentially threatened with extinction" (CITES 1994). While agarwood trades were under strict controls by official authorities, traders tend to locate alternative agarwood sources to meet the ever-increasing demand from the consumers. It was then discovered that several other *Aquilaria* species and members of at least another genera, *Gyrinops*, also produce agarwood. As a result, in the 13th CITES Conference meeting for Consideration of Proposals for Amendment of Appendices I and II, Indonesia as the proponent raised the issue for the inclusion of *Aquilaria* species and *Gyrinops* species in Appendix II. The new inclusion was accepted in 2005 (CITES 2004). Other international organization also heightened public awareness toward this genus. The global environmental organization, the International Union for Conservation of Nature and Natural Resources (IUCN), for example, listed nine *Aquilaria* species under the IUCN Red List of Threatened Species since 1998. At present, seven species are reportedly vulnerable while two others are critically endangered.

Since the very first species was identified in 1783, the understanding of the genus *Aquilaria* is still subject to reviewing and information collection from natural stands in the wild. On herbarium basis, misidentification often occurs for this genus. As *Aquilaria* trees in the wild are rarely seen with fruits and flowers, identification efforts can be really difficult. This problem was highlighted by The Wildlife Trade Monitoring Network (TRAFFIC) (Wyn and Anak 2010), giving example on Browne's (1955) findings of *Aquilaria* species occurrence in the state of Sarawak in Malaysia, speculating most of them as *A. malaccensis*. However, the author was unable to shed light on the species due to lack of reproductive parts. On contrary, Anderson (1980) clarified that *A. malaccensis* is rarely available in Sarawak. His observation was based only on a single herbarium specimen, which was collected in Marudi, East of Sarawak. Subsequently, Tawan (2004) reviewed the herbarium specimen and identified it as *A. beccariana*.

Current international effort focuses on combining multiple checklist datasets to reexamine the genus identity. One such effort is led by the Royal Botanic Gardens and Kew and Missouri Botanical Garden, which brought into being The Plant List in 2010. From the latest release, version 1.1, in May 2012, 21 out of 47 names were accepted (note: Tropicos recorded that the orthography for *A. banaense* was corrected to *A. banaensis* (Hô 1992)). As taxonomy solely depends on the morphological characteristics of the plant itself, probability of misidentification cannot be avoided. It will be a continuous effort to identify the correct species name and exclude the synonyms from time to time. Cooperation between experienced botanist and ecologist is required to reassessed the natural populations and review the type specimens to reduce the error of identification.

1.6 Domestication of the Genus

As the agarwood business is deemed lucrative, many individuals, entrepreneurs, and government agencies have ventured into *Aquilaria* plantations. Several species have been mass cultivated as plantation species in climate-suitable countries including *A. crassna* (widely spread), *A. malaccensis* (Peninsular Malaysia, West Indonesia, East India), *A. microcarpa* (Borneo region), *A. sinensis* (China), *A. subintegra* (South Thailand and Peninsular Malaysia), *A. filaria* (home gardens of East Indonesia), and *A. hirta* (home gardens of Peninsular Malaysia and West Indonesia) with the intention to sustainably produce agarwood in a short period of time. Interestingly, although *Aquilaria* is considered a highly out-crossing tree (Tangmitcharoen et al. 2008), yet, there is no formal records on hybrids in the field. There have been instances where individuals maintained that they breed for hybrid species; however, none has been scientifically supported. Perhaps the information is treated with discreet.

When it comes to species selection for planting, many farmers do not have the knowledge on which species to choose. They are often influenced by sellers of the planting materials. Most of the time, sellers rely on personal perception when recommending the species of superior growth and agarwood yield. These unfortunately are biased to the sellers, and most importantly are not corroborated with scientific evidence. It has been proven that different species produce different chemical compounds that make the agarwood scent unique to each species. In addition, the type of inducers also affects the outcome.

In general, one has to consider several criteria when selecting a plantation species:

(a) Trade market

The primary focus is targeted toward the trade market one wishes to penetrate. Different importing countries have different specific uses of the agarwood and preferences in species of origin. For example, consumers from the Middle East and India prefer agarwood from *A. malaccensis* source for perfumery and religious traditions; China regards *A. sinensis* agarwood as having medicinal properties; Korea acknowledges only *A. malaccensis* as true agarwood; and

Japan prefers agarwood from *A. crassna* for meditation purposes. However, other *Aquilaria* species possess a smaller-scale demand in perfumery, ornamental carvings, personal accessories, and cosmetics.

(b) Species adaptability

Although *Aquilaria* species are from the tropical regions, there are still possibilities that exotic species brought into as plantation species may not survive in the new planting environment. Most of the local planters cultivate native species to ensure low percentage of mortality. Furthermore, there were informal reports claiming that trees from other regions, which was proven prone to agarwood formation, actually did not produce the expected agarwood quality even after the same inoculation treatment was carried out. This could be explained by the phenotypic changes occurring in the tree, influencing it to be non-susceptible toward agarwood formation. It was suggested that such risk can be reduced by matching the seed source with similar ecological conditions to the planting site.

(c) Availability of planting resources

Like many other agriculture plants, *Aquilaria* flowers and fruits almost annually. However, the fruiting season varies among populations and tends to be inconsistent. *Aquilaria* trees are rare in the wild, numbers of mother trees that are likely prone to agarwood formation are limited; hence, it is difficult to obtain large amount of seeds suitable for plantation purposes. To ensure that the planting resources are sufficient and provide promising end products, getting seeds and planting materials from reliable sources, such as an established seed orchard or breeding center, is much recommended; however, their availability is scarce.

(d) Establishment of R&D and silviculture practices

As planting *Aquilaria* is a long-term investment, an optimal plantation management system is required to ensure planters get to harvest before anything goes wrong. Alongside with the planting, research on a suitable inoculation is indeed a crucial step to produce agarwood. To date, most of the modern inoculation techniques focus onto only a single *Aquilaria* species; therefore, planters need to be aware of their planting material beforehand. Established planting techniques, land use management, and pest and disease controls for the species of interest are also important factors for planters to select their desired planting materials. Reducing the risk of plantation investments by involving research-associated authorities will help planters to obtain firsthand information to manage their plantations.

1.7 Description of Cultivated *Aquilaria* Species

For *Aquilaria* in general, species recognition relies much on their reproductive parts, especially the fruit and calyx characteristics (Tawan 2004). Vegetative characteristic such as the leaves shape remains subjective as phenotypic changes may occur, causing alteration due to environment adaptation. From our own experience, one cannot tell the species with certainty by relying on the leaf morphology, except for *A. hirta*, because of its distinctive puberulous leaf abaxial and leaf petiole. To aid planters in identification, the vegetative and reproductive characteristics of five commonly planted *Aquilaria* species are compiled here (Table 1.3). This description is not meant for taxonomical purposes but for field workers to help identify the species. Descriptions were from field observations and published references: *A. crassna* from Schmidt and Nguyen (2004), *A. subintegra* from Ding Hou (1964), *A. malaccensis* and *A. hirta* from Ding Hou (1960), and *A. sinensis* from Flora Reipublicae Popularis Sinicae (1991).

The simplest approach to identify the species of cultivated *Aquilaria* trees in plantations is by referring to their fruit structure, also known as capsule. Typical differences of the capsule structure include the size, the shape of the fruit apex and base, the surface texture, and the size and shape of the calyx lobe (Fig. 1.2). Therefore, species identification for common plantation species can be easily carried out during their fruiting seasons. The common flowering and fruiting

Table 1.3 Vegetative and reproductive characteristics of cultivated *Aquilaria* species

	A. crassna	*A. subintegra*	*A. malaccensis*	*A. sinensis*	*A. hirta*
Size	Big tree	Shrub	Big tree	Shrub	Small tree
Leaf shape	Elliptic	Elliptic-oblong or slightly obovate-oblong	Oblong-lanceolate, caudate-acuminate	Orbicular, elliptic-oblong, obovate	Elliptic to elliptic-oblong
Leaf texture	Leathery	Papery	Papery	Leathery	Semi-leathery,
Leaf surface	Smooth	Almost smooth	Smooth	Smooth	Smooth above, Hairy beneath
Capsule shape	Round, obovoid	Ellipsoid-oblong	Obovoid or obovoid oblong	Ovoid	Oblanceolate acute
Capsule base	Slight cuneate	Cuneate	Cuneate	Cuneate	Cuneate
Capsule apex	Round	Round	Round	Narrow or round	Sharp
Capsule texture	Heavily wrinkled	Slightly wrinkled	Smooth	White silky or smooth	Hairy
Calyx lobe	Broad-ovate or –obovate, 12–15 mm long, spreading or occasionally reflexed	Ovate-oblong, 3.5–5 mm long	Ovate-oblong, 2–3 mm long, spreading or reflexed	Spreading or occasionally reflexed	Ovate and obtuse, 2–3 mm long
Calyx tube	Bell-shaped	Bell-shaped	Bell-shaped	Bell-shaped	Cylindrical
Flower length	0.38 cm	0.75–1.1 cm	2.5 cm		2.5 cm

Fig. 1.2 Fruits of five commonly cultivated *Aquilaria* species. (**a**) *A. subintegra* capsule (*top*) is egg-shaped, generally smaller compared to *A. crassna* (*bottom*), which is spherical and wrinkly, while both have broad calyx lobes pointing toward the capsule base; (**b**) *A. malaccensis* is round at the apex while slender at the base, with tiny calyx lobes recurving outward; (**c**) *A. hirta* is a bit flattened, elongated, and pointy at the apex, covered with fine hair; and (**d**) *A. sinensis* is diamond-shaped or sometimes oval with a narrow base, with calyx lobe similar to *A. crassna*, only narrower. *Arrows* point to calyx lobes. Photos: (**b**) and (**c**) Salleh Endot

Fig. 1.3 Inflorescence of several *Aquilaria* species. (**a**) *A. malaccensis* is bell-shaped and yellowish, (**b**) *A. hirta* is cylindrical, long, and fine haired, and (**c**) *A. sinensis* is bell-shaped and light green and has long petals. Photo: (**c**) Y. Yang

period for *Aquilaria* trees is between May and August (Fig. 1.3). Eventually, fruiting frequency is greatly influenced by the climate and geographical differences where the tree is populated (Soehartono and Newton 2001). *Aquilaria* fruit produces 1–2 seeds in a capsule, and seeds are important resources for cultivation (Fig. 1.4). *Aquilaria* seeds are classified as recalcitrant seeds, which are very sensitive to desiccation and cannot be kept for a long period. Under room temperature, the storage period for a successful germination is at maximum 3 days, after which the germination rate drops. When kept under −10 °C, it can be stored for a maximum of 25 days with promising survival rate (Aroonrungsikul et al. 2009; Aroonrungsikul and Wongsatain 2010; Saikia and Khan 2012; Tabin and Shrivastava 2014).

Planting materials are often young saplings ranging from 6 months to 2 years old. Species identification on young saplings is a difficult task as reproductive parts are absent. In such scenario, effective identification can only be carried out through leaf morphology. Except for *A. hirta* which has hairy leaves, all the remaining com-

Fig. 1.4 Capsules and seeds of five commonly cultivated *Aquilaria* species. (**a**) Capsule of *A. subintegra* (*left*) and *A. crassna* (*right*); the latter is round in shape, while the former is oblong with a cuneate or triangle base (*red arrow*); (**b**) seed of *A. crassna* (*right*) is reddish brown, with a wrinkled appendage, which is clearly seen in the previous photo (*yellow arrow*) and is bigger than *A. subintegra* seed (*left*); (**c**) seeds of *A. sinensis* have long and slender appendages; and (**d**) *A. malaccensis* seed is small and round and has a reddish short appendage. Capsules (**a**) and (**d**) are on the same scale. Photo: (**d**) Salleh Endot

monly cultivated *Aquilaria* species have similar leaf characteristics at young age. A study on the vegetative description of three *Aquilaria* species in Malaysia was carried out by describing distinctive vegetative characteristics of the saplings. Although it could be done, *Aquilaria* saplings identification remains a challenging job for inexperienced individuals (Lee et al. 2013). At present, identification efforts can be supported using molecular approaches such as DNA markers. Unlike conventional identification, molecular-based identification does not require a professional or experienced personal to identify tree morphology. Armed with simple laboratory procedures and basic facilities, which are available at any descent research institutions, species identification can be accomplished in a rapid manner and most importantly with confidence. A well-optimized molecular technique can be an effective tool to ascertain species identity and potentially contributes in breeding programs for *Aquilaria* (Lee et al. 2011).

In Indonesia, *Aquilaria* planting is carried out at a small to medium scale, and the trees of choice are "filaria" and "malakensis." However, this grouping does not indicate a specific species; rather it was a general trade name that distinguishes agarwood from the eastern and western regions of Indonesia, respectively, and may include *Gyrinops* species. *Gyrinops* is a closely related agarwood-producing tree under the same family that populates the same region as the *Aquilaria* (Takeuchi and Golman 2002). Planters could misidentify and wrongly reported their planting materials to the authorities, thus causing incorrect evaluation of the preferred plantation species in the region. Due to the ambiguity of the planted species, we have excluded them from this section.

1.8 Brief Notes on *Gyrinops*, a Closely Related Genus

Together under the same subfamily Aquilariodeae, the genus *Gyrinops* is the nearest tree to *Aquilaria*. These two genera are taxonomically different only by the number of stamens: *Aquilaria* has eight to twelve stamens, while *Gyrinops* has five. The first *Gyrinops* was recorded in the year 1791 by Joseph Gaertner; *Gyrinops walla* was the type specimen.

With a total of nine accepted species name under the genus *Gyrinops*, seven were found distributed in the southeast of Indomalesia regions, namely, Indonesia and Papua New Guinea (Table 1.4). Those that are endemic to Indonesia are *G. decipiens*, *G. moluccana*, *G. podocarpa*, and *G. versteegii*, while *G. caudata* is endemic to Papua New Guinea. Two other species, *G. ledermannii* and *G. salicifolia*, are found in both countries. Further north in the Indomalesia region, *G. vidalii* has been reported endemic to Laos and *G. walla* to Sri Lanka.

Similar to *Aquilaria*, the *Gyrinops* is also classified as agarwood-producing trees. To date, only three species are known agarwood producers: *G. ledermannii*, *G. versteegii*, and *G. walla* (Hou 1960; Ng et al. 1997; Compton and Zich 2002; Subasinghe et al. 2012). The genus *Gyrinops* was listed under Appendix II of CITES in 2005 together with the genus *Aquilaria* as a conservation effort; however, it was not included in the IUCN Red List perhaps because the exploitation of its natural population was not as alarming as *Aquilaria*.

In Papua New Guinea, where *A. filaria* and *G. ledermannii* can both be found, identification is compounded by them sharing the same morphology. Careful examination, however, revealed that *Gyrinops* contains the same number of stamens as the petals, while *Aquilaria* has twice the number of stamens to the petals. In terms of agarwood, no one genus is superior to the other. *Gyrinops* plantations were initiated as early as 1998 without prior information on its taxonomy. A thorough taxonomy study was then carried out in 2001 where the plantation species was identified as *G. ledermannii* (Subasinghe et al. 2012).

Because of very similar characteristics between the two genera, the idea to consolidate *Aquilaria* and *Gyrinops* became apparent. Hailier (1922) and Ding Hou (1960) acknowledged that both genera are highly similar in taxonomical characteristics, and the difference in number of stamens should not be the major factor for

Table 1.4 Accepted names and distribution of *Gyrinops* species

No.	Species name	Basionyms and Synonyms	Year first reported	Distribution
1	*Gyrinops caudata* (Gilg) Domke	*Aquilaria caudata* *Brachythalamus caudatus* *Gyrinops audate*	1932	Papua New Guinea
2	*Gyrinops decipiens* Ding Hou	–	1960	Indonesia
3	*Gyrinops ledermannii* Domke	–	1932	Papua New Guinea Indonesia
4	*Gyrinops moluccana* (Miq.) Baill	*Aquilaria moluccana* *Lachnolepis moluccana*	1946	Indonesia
5	*Gyrinops podocarpa* (Gilg) Domke	*Aquilaria podocarpus* *Brachythalamus podocarpus* *Gyrinops ledermannii* *Gyrinops podocarpus*	1932	Indonesia
6	*Gyrinops salicifolia* Ridl.	*Gyrinopsis salicifolia*	1916	Indonesia Papua New Guinea
7	*Gyrinops versteegii* (Gilg) Domke	*Aquilaria versteegii* *Brachythalamus versteegii*	1932	Indonesia
8	*Gyrinops vidalii* P.H.Hô	–	1987	Laos
9	*Gyrinops walla* Gaertn.	*Aquilaria walla*	1791	Sri Lanka

According to Version 1.1 of The Plant List (2013), http://www.theplantlist.org, assessed on 26 January 2016
– not available

natural segregation. Eurlings and Gravendeel (2005) conducted a phylogenetic analysis on both genera by comparing morphological characteristics such as the shape of the calyx tube, lobes and fruits, presence or absence of leaf indumenta, and prominence of leaf venation. These attributes seem to correlate better with molecular clades formed, rather than the number of stamens, which is highly homoplasious. With further research, it is perceivable that *Gyrinops* may be reduced into *Aquilaria* synonym as it can reflect natural resemblances better.

1.9 Conclusions

As an endangered genus, the understanding of *Aquilaria* botany and taxonomy is an essential step toward conservation in the wild. With its natural population spreading across more than ten countries, further deteriorated by its scarcity in numbers and thus being threatened to extinction, the work for conservation must be carried out with full cooperation from all authorities. New efforts to review *Aquilaria* taxonomy and systematics could be initiated by recollecting specimens of several species, starting with those that have not been updated recently. An example are the endemic species of the Philippines, whereby the latest record is from 1946. Furthermore, by taking advantage of molecular data, it may aid in taxonomy work such as in the classification of *Aquilaria* species, identification of species origin of agarwood products, and resolution of issues regarding species affinities between *Aquilaria* and *Gyrinops*.

References

Anderson JAR. A checklist of the trees of Sarawak. Kuching: Forest Department, Sarawak; 1980.

Angiosperm Phylogeny Group. An ordinal classification for the families of flowering plants. Ann Mo Bot Gard. 1998;85(4):531–53.

Aroonrungsikul C, Karwsuwan U, Wongsatain W. Preliminary study of agarwood (*Aquilaria crassna*) fruit and seed storage. Proceedings of the 47th Kasetsart University Annual Conference, Kasetsart, 17–20 Mar 2009. Subject: Plants. Kasetsart University; 2009. p. 281–8.

Aroonrungsikul C, Wongsatain W. Seed quality of some cloning of agarwood (*Aquilaria crassna*) in Trad Province area. Proceedings of the 48th Kasetsart University Annual Conference, Kasetsart, 3–5 Mar 2010. Subject: Plants. Kasetsart University; 2010.

Browne FG. Forest trees of Sarawak and Brunei. Kuching: GPO; 1955. p. 370.

Compton JGS, Zich FA. *Gyrinops ledermannii* (Thymelaeaceae), being an agarwood-producing species prompts call for further examination of taxonomic implications in the generic delimitation between *Aquilaria* and *Gyrinops*. Flora Malesiana Bull. 2002;13(1):61–5.

CITES. Convention on International Trade in Endangered Species of Wild Fauna and Flora. Consideration of proposals for amendment of Appendices I and II- *Aquilaria* spp. and *Gyrinops* spp. Thirteenth meeting of the Conference of the Parties, Bangkok, 2–14 Oct 2004.

CITES. Convention on International Trade in Endangered Species of Wild Fauna and Flora. Resolutions of the Conference of the Parties. Ninth meeting of the Conference of the Parties, Fort Lauderdale, 7–18 Nov 1994.

Cronquist A. The evolution and classification of flowering plants. 2nd ed. Bronx: The New York Botanical Garden; 1988.

Domke W. Untersuchungen über die systematische und geographische Gliederung der Thymelaeaceen. Bibliotheca Botanische. 1934;27:1–151.

Don G. A general history of the dichlamydeous plants, vol. 2. London: J.G. and F. Rivington; 1832. 4.

Duke JA. Duke's handbook of medicinal plants of the Bible. Boca Raton: CRC Press; 2007.

Eurlings MCM, Gravendeel B. *Trn*L-*trn*F sequence data imply paraphyly of *Aquilaria* and *Gyrinops* (Thymelaeaceae) and provide new perspectives for agarwood identification. Plant Syst Evol. 2005;254:1–12.

Flora Reipublicae Popularis Sinicae. 1991. [http://www.eflora.cn]. Access date 3 Aug 2015.

Gilg E. Thymelaeaceae. In: Engler A, Prantl K, editors. Die Natürlichen Pflanzenfamilien III, 6a. Leipzig: Engelman; 1894. p. 216–45.

Hailier JG. Beiträge zur Kenntnis der Thymelaeaceen und ihrer natüralichen Umgrenz-ungen. Meded Rijaks Herb. 1922;44:1–31.

Hamilton F. Chapter 11. A commentary on the second book of the herbarium amboinense. In: Memoirs of the Wernerian Natural History Society, vol. 6. Edinburgh/London: Adam Black and Longman; 1836. p. 268–333.

Herber BE. Pollen morphology of the Thymelaeaceae in relation to its taxonomy. Plant Syst Evol. 2002;232:107–21.

Herber BE. Thymelaeaceae. In: Kubitzki K, editor. The families and genera of vascular plants, Flowering plants dicotyledons (Malvales, Capparales and Non betalain Caryophyllales), vol. 5. Berlin: Springer-Verlag Berlin Heidelberg; 2003. p. 373–438.

Heywood VH. Flowering plants of the world. London: B. T. Batsford Ltd; 1993.

Hô PH. *Aquilaria* (Thymelaeaceae). In: Morat P, editor. Flora of Cambodia, Laos and Vietnam, vol. 26. Parisin: Museum National d'Histoire Naturelle; 1992. p. 51.

Hou D. Thymelaeaceae. In: Van Steenis CGGJ, editor. Flora malesiana series I, vol. 6. Netherlands: Wolters-Noordhoff Publishing; 1960. p. 1–48.

Hou D. Notes on some Asiatic species of *Aquilaria* (Thymelaceae). Blumea. 1964;12(2):285–8.

Hutchinson J. The genera of flowering plants, Dicotyledons, vol. 2. Oxford: Clarendon; 1967.

Ito M, Honda G. Taxonomical identification of agarwood-producing species. Nat Med. 2005;59:104–12.

Kiet LC, Kessler PJA, Eurlings MCM. A new species of *Aquilaria* (Thymelaeaceae) in Vietnam. Blumea J Plant Taxonomy Plant Geogr. 2005;50(1):135–41.

Lee SY, Weber J, Mohamed R. Genetic variation and molecular authentication of selected *Aquilaria* species from natural populations in Malaysia using RAPD and SCAR markers. Asian J Plant Sci. 2011;10(3):202–11.

Lee SY, Faridah-Hanum I, Mohamed R. Vegetative description of three *Aquilaria* (Thymelaeaceae) saplings in Malaysia. Pertanika J Trop Agric Sci. 2013;36(S):287–94.

Li T. Nguyen Cochinchina: southern Vietnam in the seventeenth and eighteenth centuries, vol. 23. Ithaca: Southeast Asia Program Publications; 1998.

Ng LT, Chang YS, Kadir AA. A review on agar (gaharu) producing *Aquilaria* species. J Trop Forest Prod. 1997;2(2):272–85.

Rautenbach M. *Gnidia* L. (Thymelaeaceae) is not monophyletic: taxonomic implications for *Gnidia* and its relatives in Thymelaeoideae, Doctoral dissertation, University of Johannesburg; 2008.

Ridley HN. Garu and chandan. In Royal Asiatic Society of Great Britain and Ireland Straits Branch. J Straits Branch R Asia Soc. 1901;35–36:73–82.

Royle JF. Illustrations of the botany and other branches of the natural history of the Himalayan mountains and of the flora of Cashmere, vol. 1. London: WM. H. Allen and Company; 1839.

Saikia P, Khan ML. Phenology, seed biology and seedling survival and growth of *Aquilaria malaccensis*: a highly exploited and red listed tree species of North East India. Indian Forester. 2012;138(3):289–95.

Schmidt LH, Nguyen XL. *Aquilaria crassna* Pierre. Seed Leaflet. 2004;100.

Society for the Diffusion of Useful Knowledge. The Penny Cyclopaedia of the Society for the Diffusion of Useful Knowledge, vol. 9. London: Charles Knight and Company; 1838.

Soehartono T, Newton AC. Reproductive ecology of *Aquilaria* spp. in Indonesia. For Ecol Manage. 2001;152(1):59–71.

Subasinghe SMCUP, Hettiarachchi DS, Rathnamalala E. Agarwood-type resin from *Gyrinops walla* Gaertn: a new discovery. J Trop Forestry Environ. 2012;2(2):43–8.

Tabin T, Shrivastava K. Factors affecting seed germination and establishment of critically endangered *Aquilaria malaccensis* (Thymelaeaceae). Asian J Plant Sci Res. 2014;4(6):41–6.

Takeuchi W, Golman M. The identity of Eaglewood (*Gyrinops*, Thymelaeaceae), a new economic resource for Papua New Guinea. SIDA Contributions Botany. 2002;20:261–7.

Tangmitcharoen S, Yongrattana P, Luangvirivasaeng V, Tasen W, Chantbep P, Saengtubtim S. Floral biology of *Aquilaria crassna* Pierre ex Lecomte. Thai J Forestry. 2008;27(2):1–13.

Tawan CS. Thymelaeaceae. In: Soepadmo E, Saw LG, Chung RCK, editors. Tree flora of Sabah and Sarawak, vol. 5. Kuala Lumpur: Forest Research Institute Malaysia; 2004. p. 133–484. 528 pp.

The Plant List. 2013. Version 1.1. [http://www.theplantlist.org/]. Access date 26 Jan 2016.

Tropicos. Missouri Botanical Garden. 2016. [http://www.tropicos.org]. Access date 27 Jan 2016.

Van der Bank M, Fay MF, Chase MW. Molecular phylogenetics of Thymelaeaceae with particular reference to African and Australian genera. Taxon. 2002;51(2):329–39.

Watt G. A dictionary of the economic products of India, vol. 1. Cambridge: Cambridge University Press; 2014.

Wyn LT, Anak NA. Wood for the trees: a review of the agarwood (Gaharu) trade in Malaysia. Traffic Southeast Asia; 2010.

Chapter 2
Wood Resources, Identification, and Utilization of Agarwood in China

Yafang Yin, Lichao Jiao, Mengyu Dong, Xiaomei Jiang, and Shujuan Zhang

Abstract Agarwood, a compound substance of xylem tissue and its inclusions, is generally formed in some genera of Thymelaeaceae (e.g., *Aquilaria*) during the tree's natural growing process. Due to the medicinal and economic values of agarwood augmented by unsustainable harvesting techniques, calls have been made by international bodies such as CITES and IUCN to conserve agarwood-producing tree species. Consequently, identification of agarwood resources becomes an important issue to provide better protection for the species. In this chapter, we review the status, i.e., forest resources, wood features, identification, and market development of agarwood (*Aquilaria*) in China. Some suggestions for the sustainable development of *Aquilaria* resources in China are also provided.

2.1 Introduction

Agarwood has a distinctive fragrance and is a precious traditional medicine and a much-sought after perfume in Asian countries such as China, Japan, India, and countries in the Middle East. During the past 30 years, there has been a surge in demand for agarwood worldwide. However, due to its slow formation, limited output, and high volume of trading, the supply of agarwood has decreased gradually, and its price is constantly increasing. For better understanding on the status of agarwood development, the distribution of *Aquilaria* resources in China, both in the wild and planted, is provided here. Additionally, we discuss identification methods of *Aquilaria* resources based on wood anatomy, DNA barcode, and chemical analysis. However, the latter is preferably used to test for agarwood quality. To identify *Aquilaria* wood at species level, based on its wood anatomical features alone, is a difficult if not impossible task. Nevertheless, the newly developed DNA barcoding

Y. Yin (✉) • L. Jiao • M. Dong • X. Jiang
Department of Wood Anatomy and Utilization, Research Institute of Wood Industry, Chinese Academy of Forestry, Beijing 100091, P.R. China
e-mail: yafang@caf.ac.cn

S. Zhang
Beijing Tian-yi-li-hua Institute of Agarwood, Beijing, P.R. China

R. Mohamed (ed.), *Agarwood*, Tropical Forestry,
DOI 10.1007/978-981-10-0833-7_2

technology might transcend these limitations and result in effective information with a high resolution. Adopting such technique for specific and rapid identification of agarwood species will help conserve its natural resources in the forest while creating a sound and orderly environment for development of agarwood market.

2.2 Distribution of Agarwood Resources

2.2.1 Global Distribution

Discussions have taken place on the scope of the plant species that can produce agarwood. It is generally recognized that agarwood is mainly produced by the Thymelaeaceae trees, which include *Aquilaria*, *Aetoxylon*, *Enkleia*, *Gyrinops*, *Gonystylus*, *Phaleria*, and *Wikstroemia*; however, not all can form agarwood (Wyn and Anak 2010). Among them, the *Aquilaria* trees are widely recognized in China as the main source for producing agarwood. In this chapter, *Aquilaria* is the main genus in focus.

The wild resources of global *Aquilaria* plants are mostly concentrated in the southeast of Asia. There are about 20 *Aquilaria* species, the most common being *A. malaccensis*, *A. crassna*, *A. microcarpa*, *A. filaria*, *A. subintegra*, and *A. beccariana*. When it comes to Eastern Asia, there is a wide distribution of *A. sinensis* in Southern China. *Aquilaria yunnanensis* is also found in the same region but its distribution is sparser. Among the countries that produce agarwood, the Philippines has the most species of *Aquilaria* and the richest variety accounting for half of the world's total. Next is Indonesia, which has one third of the total. It also has the largest distribution area of resources for *Aquilaria* trees. Malaysia and Thailand are also rich in *Aquilaria* species. By comparison, there are fewer native *Aquilaria* species in Vietnam, Myanmar, Laos, India, and China. According to the records of the International Union for Conservation of Nature (IUCN), there is also a distribution of *A. malaccensis* in Iran in West Asia.

Due to the enormous market demand for agarwood and excessive felling and low natural growth rate for *Aquilaria*, the tree distribution has significantly decreased over the past 20 years in the Asian region. Meanwhile, for meeting the increasing demands of the international agarwood market, the development of plantations has now come into research focus. At present, *Aquilaria* trees are widely cultivated in many countries with the technology for artificial agarwood induction being increasingly perfected.

2.2.2 Distribution in China

There are currently two native *Aquilaria* species in China, i.e., *A. sinensis* and *A. yunnanensis* (Cheng et al. 1992; Editorial Board of Flora of China of Chinese Academy of Sciences 1999). Between the two, *A. sinensis* is the only plant resource

Fig. 2.1 Planted *Aquilaria sinensis* in Guangdong Province

that can be used for producing agarwood for traditional medicine in China (Fig. 2.1). It has been used as genuine medicine for more than a thousand years. *Aquilaria sinensis* originates in Hainan, Guangdong, and parts of Guangxi and Yunnan provinces. At present, it is mainly produced in the Hainan Province. As early as the Tang and Song Dynasties, *A. sinensis* has been widely cultivated in the Dongguan area, where it became a local specialty. Consequently, agarwood has also been called "Guan Xiang," which means the agarwood of Dongguan. According to textual research on medicinal history, the evolution of the variety, and the origin of agarwood in "Nanyao" (the traditional medicine in South China), the sources of medicinal agarwood have been grouped into native agarwood (*A. sinensis*) and imported agarwood since the time of Ancient China (Mei et al. 2011).

2.2.2.1 Wild Resources

At present, the distribution of wild resources of *A. sinensis* occurs mainly in Guangdong, Hainan, and Guangxi provinces. Due to excessive felling in recent years, its wild resources have been greatly destroyed. *Aquilaria yunnanensis* S. C. Huang (http://db.kib.ac.cn), a new *Aquilaria* species found in China, mainly originated in Mengla Town (Youle Mountain) and Shuangjiang Town in Yunnan Province, but its distribution is mostly scattered or patchy.

Current distribution of wild resources of *Aquilaria* tree is becoming sparse in China. Guangdong and Hainan provinces, the two most renowned regions for agarwood in China, have *A. sinensis* scattered in coastal, offshore, hilly and low mountain areas, and within limits of certain natural protection zones. To guarantee the sustainable development of *Aquilaria* resources, the Dongguan Botanical Garden of Guangdong Province has started to preserve a considerable number of *A. sinensis* resources. It has collected *A. sinensis* from several provenances including six from Hainan Province, three from Guangxi Province, two from Yunnan, and 20 from Guangdong Province, where special works on seedling cultivation have been carried out. Historically, Guangxi Province was once one of the important distribution areas of wild resources of the *Aquilaria* tree. However, according to official 2011 statistics, there were only 354 wild *A. sinensis* occurring in a relatively concentrated area, e.g., Pubei County, whereas in Yunnan Province, there were only a few wild resources of *A. sinensis* and *A. yunnanensis* that were scattered in the natural protection zones.

The protection of *A. sinensis* is an important issue in China. The few remaining wild *A. sinensis* resources have even become the target of offenders who are increasingly destroying trees and illegally felling them. In 1992, the China Plant Red Data Book emphasized that the quantity of *A. sinensis* was seriously decreasing. In 1998, the IUCN listed *A. sinensis* on the list of Endangered Species of Wild Fauna and Flora. A year later, *A. sinensis* was listed in the second-class category of the National List of Local Protected Flora, issued by the Chinese government (The State Council of the People's Republic of China 1999). Consequently, all species from the genus *Aquilaria* have been listed in Appendix II of CITES (the Convention on International Trade in Endangered Species of Wild Fauna and Flora) since 2005 (http://www.cites.org/eng/app/appendices).

2.2.2.2 Plantation Resources

Since 2007, the Chinese agarwood market has experienced rapid growth. Due to their geographic advantage of being in the tropics and subtropics and because of their mild climate conditions, Guangdong, Guangxi, Hainan, Yunnan, and Fujian provinces have all started planting *Aquilaria*. At present, millions of *Aquilaria* trees are being cultivated, with a great deal of success and progress (Fig. 2.2). According to statistical data published in 2011, the total new planted area of national *Aquilaria* plantation resources from 2006 to 2010 had reached 5285 hectares. Of these, *A. sinensis* occupied 5245 hectares, accounting for 99 % of the total area (Table 2.1). The planting density of *A. sinensis* was generally 2 m × 1.5 m or 2 m × 3 m, with 1500–3000 trees per hectare.

A. crassna, which occupied 40 hectares and accounted for less than 1 % of the total, is a relatively successful exotic species. The species grew under good conditions, after being tested, which resulted in the successful accomplishment of artificial agarwood induction. Since the leaves of *A. crassna* are bitter tasting, they have the capacity of being insect resistant; they seldom suffer from plant diseases introduced by insects. In addition, some *A. subintegra* have also been introduced.

Fig. 2.2 Logs of *Aquilaria sinensis* plantation in Guangdong Province

Table 2.1 Resources of new planted *Aquilaria* plantation in China (2006–2010)

Province	Species	Area (ha)					Total area (2006–2010)	Percentage (%)
		2006	2007	2008	2009	2010		
Guangdong	*Aquilaria sinensis*	1528.4	39.7	132.2	260.4	918.5	2879.2	54. 45
Guangxi	*Aquilaria sinensis*	1.5	1.1	0	6.5	42.0	51.1	0.97
Hainan	*Aquilaria sinensis*	39.1	28.5	26.1	44.8	121.4	259.9	4.92
Fujian	*Aquilaria sinensis*	0.1	0	2.0	1.4	1.8	5.3	0.10
Yunnan	*Aquilaria sinensis*	466.7	300.0	453.3	433.3	396.7	2050	38.76
Subtotal		2035.8	369.3	613.6	746.4	1480.4	5245.5	99.24
Guangdong	*Aquilaria crassna*	40.0					40.0	0.76
Guangdong	*Aquilaria subintegra*	A few						
Total		**2075.8**	**369.3**	**613.6**	**749.4**	**1480.4**	**5285.5**	**100 %**

Note: The total distribution area of *Aquilaria* resources could be higher because only data of new planted *Aquilaria* resources collected since 2006 are presented here. Data from the Endangered Species Import and Export Management Office of the People's Republic of China (2011)

Guangdong Province has the longest history in *Aquilaria* planting in China. In 1978, Shantou City initiated plantations of *Aquilaria* while introducing *A. crassna* from the Kachin State of Myanmar. They were soon afterward introduced into Yunnan, Hainan, and Guangxi provinces. After 2006, *A. sinensis*, recognized as the native tree of the best quality has been widely cultivated throughout Guangdong Province. The transformation of an ecological noncommercial forest in the Pearl River Delta region has also become an effective ex-situ conservation measure for *A. sinensis*. Currently, the planting of *A. sinensis* in Guangdong Province is concentrated in Dongguan, Dianbai, Gaoyao, and Huazhou cities. This work gradually spreads throughout the province, where the scale of planting in Dianbai City is relatively considerable. Until the end of 2010, the total area of planted *A. sinensis* was 3867 hectares; 2000 hectares belong to plantations and the remaining to individual farmers.

The mass development of *Aquilaria* plantations in Hainan Province started with the introduction of a local development project. According to incomplete statistics in 2012, the plantation area is over 2000 hectares, with about three million trees distributed mainly in the Dingan, Danzhou, Chengmai, and Dunchang counties. At present, local areas of *Aquilaria* plantations are continually expanding in most cities and counties in the Hainan Province. With more people realizing the economic value inherent in planting *Aquilaria*, a developing pattern of corporatization and scaling is gradually taking place.

In Guangxi Province, the development of *Aquilaria* plantations started relatively late, focusing mainly on *A. sinensis*. Until 2011, the relatively concentrated new planted areas of 51 hectares were mainly found in Shangsi and the Fangcheng counties, among others. According to statistics in Yunnan Province, the plantation area of *Aquilaria* included about 733 hectares in Xishuangbanna and 67 hectares in Puer City. The main species was *A. sinensis* along with some *A. yunnanensis*.

Fujian Province is not a natural distribution zone for *Aquilaria*. However, it is currently carrying planting activities. By the end of 2011, the number of new planted *A. sinensis* trees mainly in young seedlings had reached about 50,000 in an area of 5.3 hectares, and they are mainly found in Quanzhou and Zhangzhou cities.

2.3 Wood Anatomical Characters of Agarwood

An examination of the anatomical characters of *Aquilaria* wood can be undertaken at both the macroscopic and microscopic levels. A macroscopic examination can be made with the naked eye or with the aid of a small magnifying glass. A microscopic examination requires the sectioning of a sample with an optional staining of the sections, followed by observation under a light microscope and making a comparison with reference samples (wood xylarium). The wood anatomical characteristics are described in the standard terminology of the International Association of Wood Anatomists (IAWA) (IAWA Committee 1989).

2.3.1 Macroscopic Features

When observing the macroscopic feature of *Aquilaria* wood with the naked eye or magnifier with a magnification of 10×, the following should be noted: the wood color, odor, growth rings, vessel arrangements and size, rays, and included phloem.

The wood color is yellowish white (Fig. 2.3). Once the wood is exposed to the air for a long term, its surface will turn dark (Fig. 2.4a). The wood is glossy and has a mild fragrant and sweet odor. If the xylem starts to produce agarwood, black lines or conglomerations will appear in that place. After producing more agarwood, the entire piece of wood will become black or dark brown.

Growth rings are indistinct. The wood is diffuse porous. The vessels are slightly small to medium, distinct under magnification. The size of vessels is consistent and evenly distributed in a dispersive arrangement. The axial parenchyma is absent. The rays are small to medium, very fine to slightly fine, and visible under magnification. Ripple marks and intercellular canal are absent. The included phloem is visible with the naked eye. There is an island-type pattern with uniform distribution, which is the key character when it comes to *Aquilaria* wood identification (Fig. 2.4b). However, the included phloem with pores should be distinguishable.

2.3.2 Microscopic Features

For microscopic observation, wood samples were excised into small blocks [10 mm (L) × 10 mm (R) × 10 mm (T)] with a razor blade and then softened in water at 80 °C for 5 h. Thereafter, transverse, radial, and tangential sections were cut into

Fig. 2.3 Wood stem of *Aquilaria* spp.

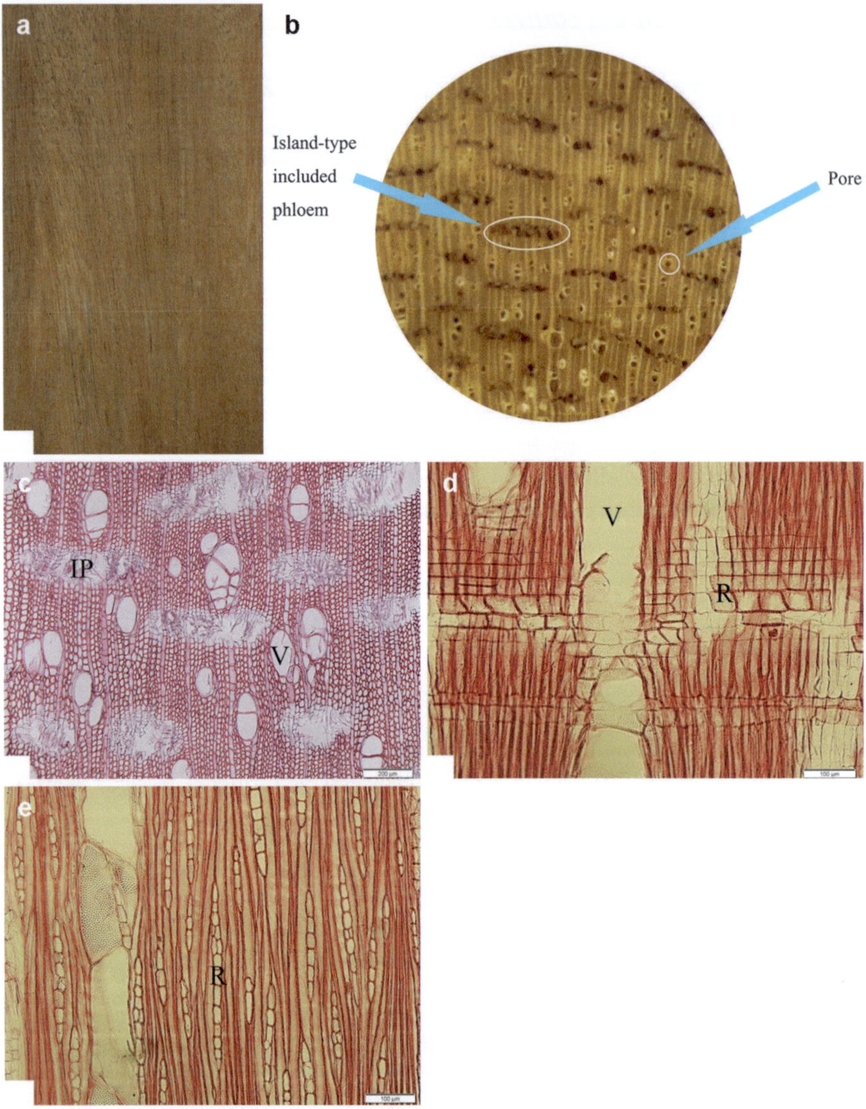

Fig. 2.4 Wood specimen (**a**), wood cross section (16×) (**b**), and anatomical features (**c–e**) of *Aquilaria sinensis*. (**c–e**) Transverse, radial, and tangential sections, respectively. Scale bars, 200 μm (**c**) and 100 μm (**d**, **e**). *IP* included phloem, *R* ray, *V* vessel

thicknesses of 15–20 μm on a sliding microtome and then observed under a light microscope (Olympus BX61, Japan) after being stained with 1 % aqueous safranine. The wood can be identified according to the features of the three wood sections. The main characteristics to be observed are the included phloem, the vessel arrangements and size, the intervessel pits, the axial parenchyma, the fibers, and the rays (Figs. 2.4c–e).

Included Phloems They are better observed in the transverse section. They are unevenly oval or elliptically long and evenly distributed.

Vessels The vessels are round or oval, sometimes partly with an angular outline. They are in radial multiples of 2–5 (mainly 2–4), occasionally with vessel clusters. They are diffuse, thin walled. Their tangential diameters are mainly 70–130 μm, with a maximum of up to 150 plus μm. They are partly storied. Tyloses are absent, with few inclusions. Helical thickenings are absent. The simple perforations are round or oval. The perforation plates mostly are slightly sloped, few slope, and parallel. The intervessel pits are alternate. The vessel-ray pits are similar to intervessel pits in size and shape.

Axial Parenchyma The axial parenchyma cells are scarce and paratracheal. They have end wall thickenings that are distinct. Gums and crystals are absent. There are few inclusions.

Fibers The fibers have thin to very thin walls, with more bordered pits; they are distinct with round or oval pit apertures included or extended, lenticular, and crack shaped, usually circular or X shaped.

Rays The rays are nonstoried, 5–14/mm, mostly uniseriate with occasional biseriate rays. They are 3–12 cells in height. They are mostly uniseriate rays with procumbent, square, and upright cells mixed throughout the ray. They have body ray cells procumbent with one row of occasional square marginal cells. The height of the upright or square ray cells is greater than that of the procumbent ray cells, with the latter being rectangular. The inclusions are usually visible. Crystals are absent. The thickening in the end walls is slightly distinct. Pits are indistinct in the horizontal walls (Figs. 2.4c–e).

2.4 Identification Methods for Agarwood

2.4.1 Wood Anatomy

Wood identification is essential in the context of timber trade, to combat illegal logging, for wood certification, and for forensic know-how. In order to achieve a definitive identification of *Aquilaria* wood, macroscopic and microscopic examinations are required in most instances. The observation of wood anatomy generally facilitates an identification at genus level (*Aquilaria* spp.), since the wood characteristics tend to be very well conserved within the genera.

However, there are other woods, belonging to mainly the genera *Memecylon* (Melastomataceae) (Fig. 2.5a, b) and *Strychnos* (Loganiaceae) (Fig. 2.6a, b), that are similar to *Aquilaria* wood. Both have included phloem, so they have been sold under the guise of *Aquilaria* wood on the wood market in the Guangdong, Guangxi, and Yunnan provinces. Fortunately, there are still some differences in their wood anatomical

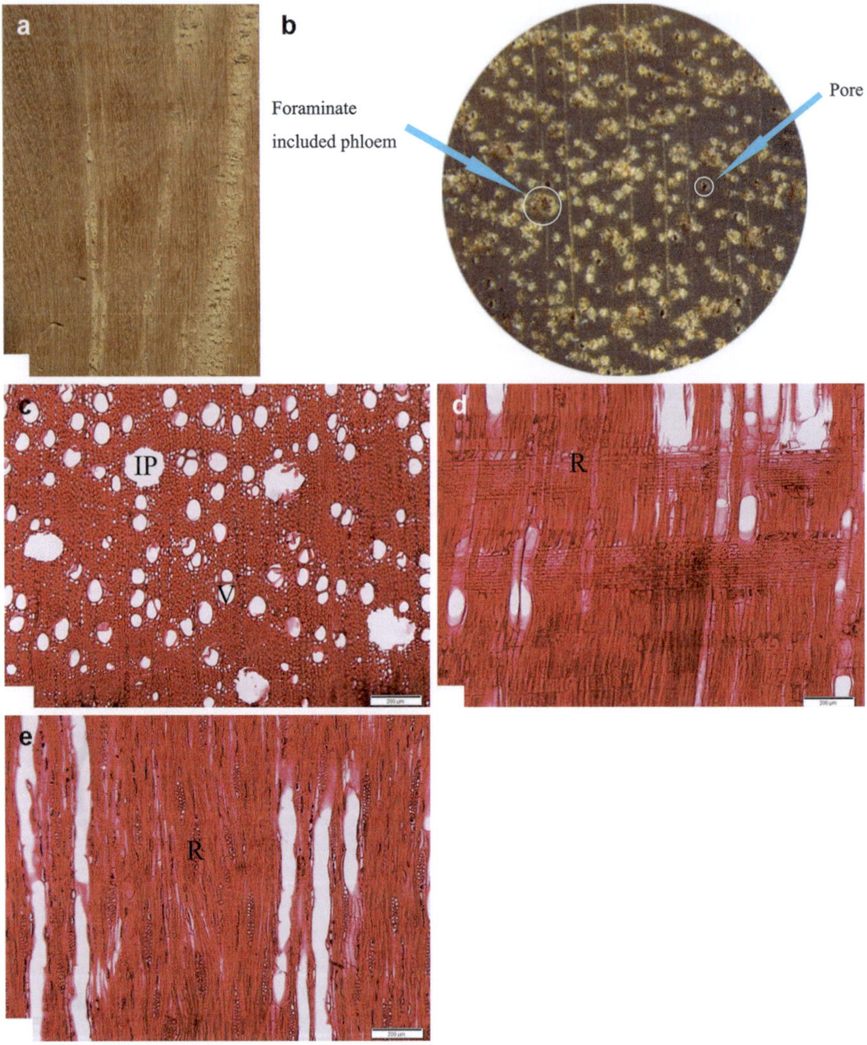

Fig. 2.5 Wood sample (**a**), wood cross section (16×) (**b**), and anatomical features (**c**–**e**) of *Memecylon* spp. (**c**–**e**) Transverse, radial, and tangential sections, respectively. Scale bars, 200 μm (**c**) and 100 μm (**d**, **e**). *IP* included phloem, *R* ray, *V* vessel

features, when comparing them to *Aquilaria* wood (Miles 1978; Cheng et al. 1992; Liu et al. 2008). The main characteristics of *Memecylon* and *Strychnos* are as follows.

2.4.1.1 *Memecylon*

Vessels are mainly solitary and partly in radial multiples. Axial parenchyma is a winged aliform, confluent-like, vasicentric, and diffuse in aggregates. Most rays are uniseriate; the multiseriate rays consist of 2–4 cells in width. The height of the upright or square

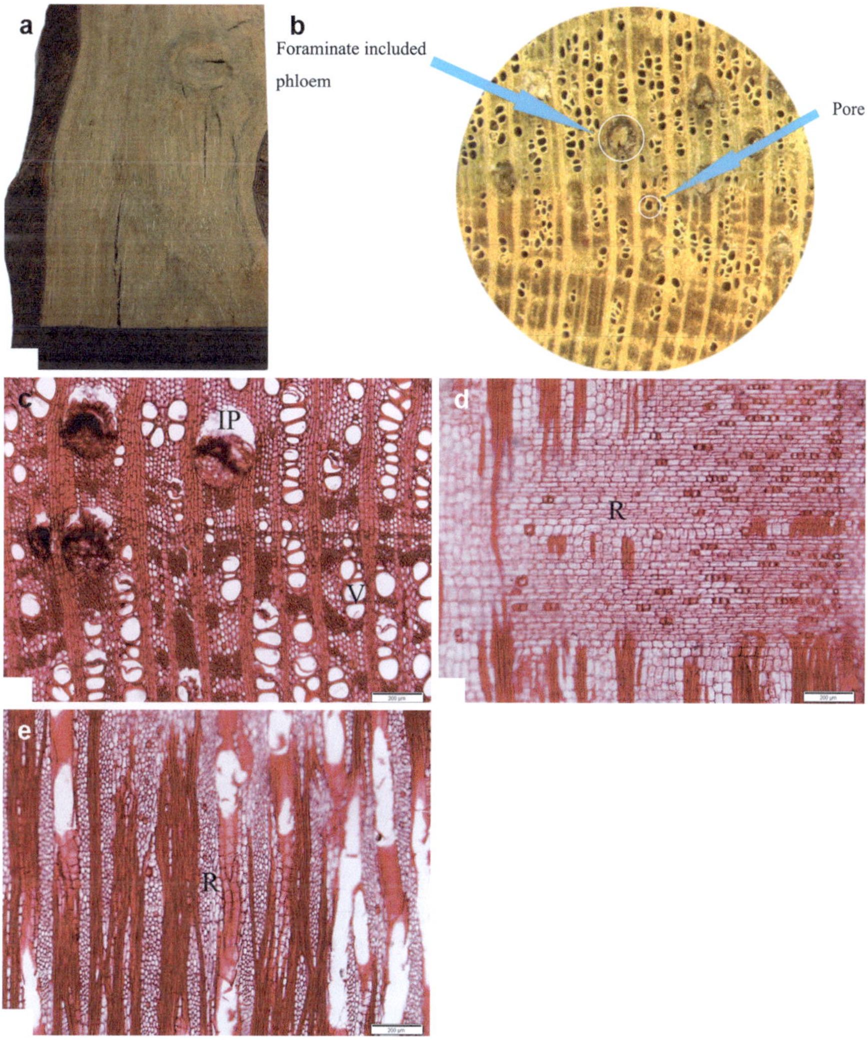

Fig. 2.6 Wood sample (**a**), wood cross section (16×) (**b**), and anatomical features (**c**–**e**) of *Strychnos* spp. (**c**–**e**) Transverse, radial, and tangential sections, respectively. Scale bars, 200 μm (**c**) and 100 μm (**d**, **e**). *IP* included phloem; *R* ray, *V* vessel

ray cells is greater than that of the procumbent ray cells. The body ray cells are procumbent with two to several rows of upright and/or square marginal cells (Fig. 2.5c–e).

2.4.1.2 *Strychnos*

The vessels are mainly solitary, with partially multiple radials up to 10 plus cells. The parenchyma is in marginal or seemingly marginal bands. The banded parenchymas are 2–8 cells in width. The larger rays are normally two to eight seriate.

Uniseriate rays are absent or extremely rare. The body ray cells are procumbent with two to several rows of upright and/or square marginal cells (Fig. 2.6c–e).

There are currently some other kinds of woods that are being sold under the guise of *Aquilaria* wood on the local market in China including *Dalbergia* spp., *Terminalia* spp., *Thuja* spp., and even *Cocos nucifera*, among others. The differences between their anatomical features and those of *Aquilaria* wood are huge. However, after treatment with resin and essential oils, these woods become dark and submerged, and the odor is similar to agarwood. Therefore, based on appearance only, it is difficult to distinguish between genuine and counterfeit agarwood. When necessary, samples should be sent to professional institutions for identification purposes.

2.4.2 DNA Barcode

DNA barcoding is a genetic approach based on a short DNA sequence from a standard part of a genome. Differences in the nucleotide sequence, specifically in targeted DNA regions, are utilized for species identification. Currently, the chloroplast genome regions, such as *rbc*L, *mat*K, *trn*L-*trn*F, and *psb*A-*trn*H, and the nuclear ribosomal DNA internal transcribed spacer (ITS), have emerged as good candidates for plant DNA barcoding (Kress et al. 2005; Gonzalez et al. 2009).

There are only a few studies on identification of *A. sinensis* based on DNA barcoding of fresh plant tissues (Eurlings and Gravendeel 2005). The main obstacle is because the available and identified material is usually dried and has been stored for a long time. To increase our chance in getting relatively intact DNA, we modified a commercial DNA extraction kit protocol for the use with *Aquilaria* samples derived from dried wood and a xylarium specimen stored for 39 years. The yielded DNA samples were successfully used in PCR amplifications. However, PCR was not successful when using DNA samples extracted from the intra-wood that were dried at 120 °C for 10 days or from the intra-wood of the xylarium specimen. Extracting DNA from dry wood treated at high temperatures or when wood has been stored for a long time is challenging (Schlumbaum et al. 2008; Rachmayanti et al. 2009; Jiao et al. 2012) because the wood DNA had extremely degraded. Nonetheless, we were able to differentiate *A. sinensis* from other species in the *Aquilaria* genus with the aid of genetic loci differences and phylogenetic analysis (Jiao et al. 2014) (Fig. 2.7).

For the *trn*L-*trn*F region, a strict consensus tree is shown, with bootstrap supports indicated at the nodes (Fig. 2.7a). The maximum parsimony (MP) analysis showed that sequences obtained from fresh wood, dried wood, wood xylarium, and leaves clustered together with *A. sinensis* sequences from the GenBank, supported by a bootstrap value of 74 % (Fig. 2.7a). Meanwhile for the ITS region, the total aligned length was 266 characters, of which 29 (10.9 %) were variable characters and 12 (4.5 %) were parsimony informative. The topology of the bootstrap 50 % majority-rule consensus tree of the MP analysis showed that sequences from fresh wood, dried wood, wood xylarium, and leaves clustered together in polytomy

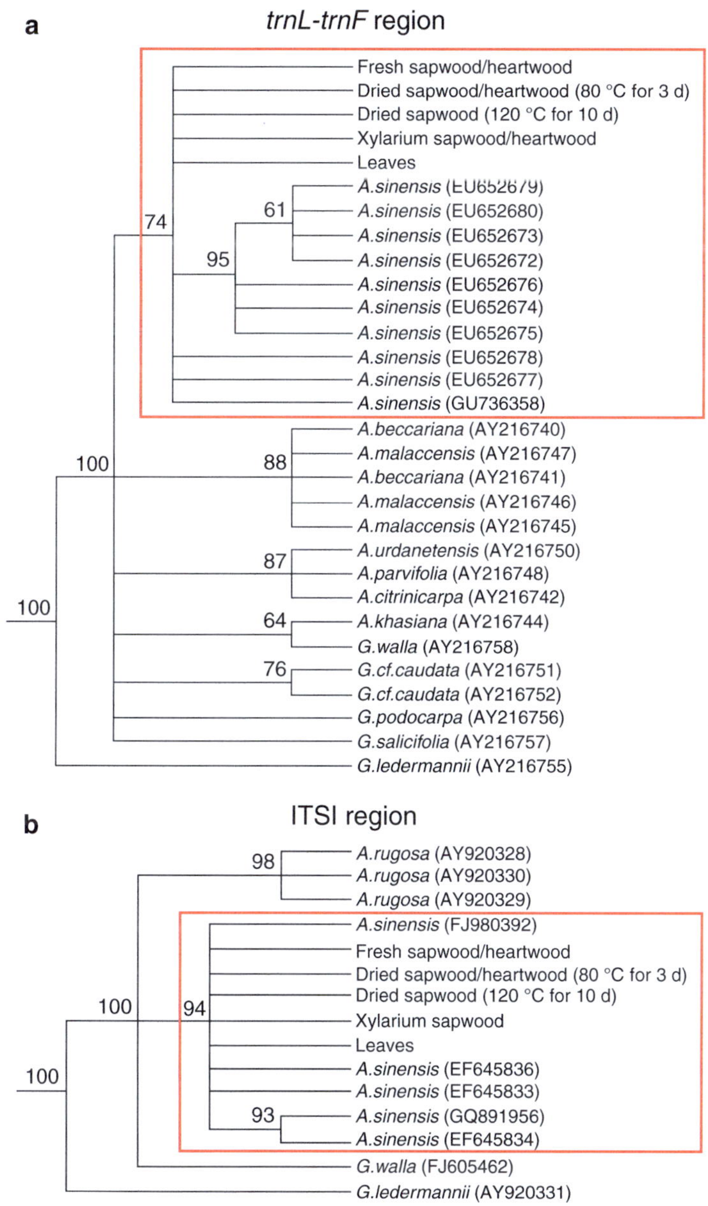

Fig. 2.7 A 50 % majority-rule consensus tree obtained with the MP method, based on (**a**) the *trn*L-*trn*F region and (**b**) the ITS1 region. The branches indicate the percentage of bootstrap values calculated from 1000 bootstrap replicates. *A Aquilaria*, *G Gyrinops* (Source: Jiao et al. (2014))

with *A. sinensis* sequences deposited in the GenBank, with 94 % bootstrap support (Fig. 2.7b). These results indicate the potential use of DNA barcoding in *Aquilaria* wood identification at species level. However, there is a need to improve the process of DNA extraction from severely degraded wood such as heartwood that has been dried and stored for a long period of time. DNA barcoding technology is an effective, feasible, and promising wood identification tool for timber trade control and forensic work, supporting conventional wood anatomy approach.

2.4.3 Chemical Analysis

As explained above, the main identification method for the determination of *Aquilaria* wood is to analyze and recognize the key features, based on macrostructure, microstructure, and DNA sequences. However, because agarwood is a compound substance, identifying the timber species accurately is only the first significant step.

At present, the medicine administration agencies in China are using China Pharmacopoeia (2010 edition) as a reference when implementing a technical standard for identifying agarwood. Apart from testing the anatomical features of agarwood, its ethanol-soluble extract content must also be tested using the hot-dipping method. A value of not less than 10.0 % is deemed acceptable. Thin-layer chromatography (TLC) is also an important method for testing agarwood in the market. With this method, a solution is prepared with 0.5 g sample, and its chromatography is compared to that of the control by checking the corresponding location of the compound's fluorescence under an ultraviolet lamp.

In the Chinese domestic market, imported agarwood is often mixed with native, causing huge differences in the price of different agarwood types or classes, and may result in the frequent occurrences of fake agarwood. In some cases, premature agarwood without inclusions can be made into handcrafts, e.g., beads and carvings, as well as medicinal raw materials, after it is given suitable treatment such as dyeing, smearing, or dipping into agarwood oil, rosin, or other chemical perfumes. This is causing a major concern, as an accurate judgment on timber identification based on scientific analysis will not be achieved by depending on a particular method only such as the ethanol-soluble extract content. In other words, the existing testing technology cannot fully meet practical demands in quality supervision and identification of agarwood.

Up to now, there has not been a testing standard or a grade standard that is internationally acknowledged, although there are some research findings that could be applicable depending on the situation. For example, agarwood has been classified into five grades according to its odor, e.g., sweet, acidic, peppery, salty, and bitter (Morita 1992), or if agarwood oils into grades known as Ultra A, B, C, and D (Heuveling van Beek and Phillips 1999). Others focus on the chemical composition analysis (Ishihara et al. 1992; Konishi et al. 2002). Over the years, Chinese researchers in the fields of wood science, medicobotany, molecular biology, and other related fields in plant sciences, have done a great deal of studies on the subject of agarwood, using many technical methods, such as the infrared spectroscopy (IRS), thermogravimetry (TG), mass spectrometry (MS), and gas chromatography-mass spectrometry (GC-MS). For example, a detailed analysis on chemical contents such as sesquiterpenes and others in agarwood has been conducted by GC-MS (Yang 1998; Lin and Qi 2000).

Recently, the Research Institute of Wood Industry at the Chinese Academy of Forestry has initiated a research work to establish a technology for identifying agarwood, with the aim of setting up an agarwood standard for the industry in China. The identification process follows a series of steps: (1) to observe and analyze the macrostructural and microstructural features, (2) to detect the alcohol-soluble extract content and conduct TLC analysis following requirements of the China Pharmacopoeia, and (3) to employ GC-MS for detection of major chemical contents, e.g., aromatic compounds, sesquiterpenoids, and 2-(2-phenylethyl) chromones, among other compounds that are contained in agarwood. The different chemical contents in agarwood could be an important index for measuring the various grades of agarwood. Additionally, the performance of thermal decomposition at different temperatures can be tested using the TG method, while any heavy metals, such as Pb and Hg, in the sample can be detected with inductively coupled plasma mass spectrometry (ICP-MS). Using Fourier transform infrared (FT-IR) analysis, the location and scope of the characteristic absorption peak of its primary chemical contents can be determined. All these sophisticated detection systems are conducive toward realization of a rapid-testing technology for agarwood identification and quality assurance.

2.5 Utilization and Trade of Agarwood in China

Agarwood is traded in various ways, including as a whole plant piece, wood blocks, wood chips, oil, and even waste powder. In Asia, these are widely used in perfumes, medicines, works of art, and others. The annual trading volume of agarwood is approximately several hundred tonnes, with its trading scale reaching millions of dollars. There are two major terminal markets in the international trade of agarwood. They are northeast Asia (Taiwan, Japan and South Korea) and West Asia (the Middle East). As a very important trading hub, Singapore has been the largest trading market for the import and export trade of agarwood. The import volume alone of agarwood wood chips in the Middle East rapidly increased from 56 tonnes in 2004 to 162 tonnes in 2007, an increase of 300 % within 4 years. Taiwan and Hong Kong in China have also been major importing regions. Part of these imports were transhipped and sold to Mainland China. As the biggest terminal market, the import volume into Taiwan increased to 402 tonnes from 1995 to 1997.

The very first book in the world concerning agarwood trading came from official Chinese Customs in AD 1200. According to the records, agarwood trading activities were mainly conducted in Borneo, Sumatra in Indonesia, the Malay Peninsula, and Cambodia. In addition, agarwood history goes back more than a thousand years ago as an application in traditional herbal medicine in China. Up to now, many Chinese herbal medicines have agarwood as an ingredient, such as the "Bawei Chenxiang San," which contains eight kinds of medicinal herbs, with agarwood as its major ingredient. Furthermore, the perfume from agarwood was the exclusive favorite of emperors, high officials, and noble lords in ancient China while also being one of the psychological dependencies of the literati class, such as scholars and bureaucrats. In the Song Dynasty, agarwood was recognized as the King of Perfumes. This was when the culture of perfume reached its peak. The custom of using perfume was intensely promoted afterwards. In the Ming Dynasty, the price of agarwood became particularly high. An old saying goes – "An inch of agarwood is worth an inch of gold." However, the fall of the late Qing Dynasty discouraged people from using agarwood perfume so it was gradually put to the sidelines and forgotten. It was not until modern times, together with the economic and cultural development of the Chinese society, that people's attention was once again drawn to agarwood, and this has contributed to the rapid development in the agarwood market in recent decades.

As the technologies for the cultivation of *Aquilaria* plantations and artificial agarwood induction are perfected progressively, a fundamental condition is provided for the agarwood market in China. At present, some enterprises in Guangdong, Guangxi, Hainan, and Yunnan provinces are conducting beneficiary exploration for industrialization development of agarwood by developing various agarwood products, including tea, oil, alcohol, bracelets, herbal products, powders, ointments, and incense, among others. Diversification of agarwood products and market-driven factors had increased the price of agarwood to more than 10-fold in 2011 from its 2007 price. In general, the market benefits of *Aquilaria* plantations are still considerable. On the other hand, China has held agarwood expos several times. For instance, the China (Dongguan) Agarwood Culture & Art Expo, which was held in 2011, had attracted 150 enterprises to the exhibition from the domestic provinces, and Southeast Asian countries, involving approximately 400,000 visitors. This expo brought in approximately RMB 300 million along with contracted projects worth approximately RMB 250 million.

The city of Putian in the Fujian Province has become the major center in China for making, collecting, and distributing domestic wood-carved handicrafts. It has also become the largest trading center for sandalwood and agarwood in the world. Up to now, native Fujian agarwood has been mainly used for producing wood-carved handicrafts, such as Buddha figures, hanging decorations, and Buddha beads, while the waste is used in the making of spices. Just in the first half of 2011, the city of Putian made use of about 2.75 tonnes of agarwood, reaching revenues of approximately RMB 26.75 million, with agarwood handcrafts accounting for 75 % of this.

The expanding demands of the domestic market in China have caused the volume of imported agarwood to increase from year to year. In 2013, the import volume reached 81.42 tonnes (Table 2.2), mainly in the form of wood blocks, wood chips, medicinal herbs, and pure essential oils. These were shipped mostly from Singapore to China; the tree sources were *A. malaccensis*, *A. crassna*, and *A. filaria*. Almost all of the imported agarwood was sold to the city of Putian, while the exported agarwood products, mainly in the form of Chinese herbal medicines and wood, were exported to Japan, Malaysia, Thailand, and the USA; the tree sources were *A. sinensis* and *A. malaccensis*. Due to the expansion of domestic demands and the limitation of agarwood resources in China, the export volume of agarwood in 2013 was a mere 40.2 kg (Table 2.2).

Table 2.2 Import and export status of *Aquilaria* products of China (2010–2013)

	Export (kg)				Import (kg)			
Species	2010	2011	2012	2013	2010	2011	2012	2013
Aquilaria sinensis	107.0	175.2	58.6					
Aquilaria malaccensis	58.8	77.6	167.2	40.2	7157.4	504.5	14,130.3	17,996.7
Aquilaria crassna					5287.2	9900.0	2.5	29.5
Aquilaria filaria					60.0	13,243.0	6756.6	62,406.0
Aquilaria spp.					27.6	2.2	5.0	998.0
Total	**165.8**	**252.8**	**225.8**	**40.2**	**12,532.2**	**23,649.7**	**20,894.4**	**81,420.2**

Note: Data from the Endangered Species Import and Export Management Office of the People's Republic of China

An enormous market profit is the fundamental reason for the cause of illegal felling of *Aquilaria* trees and trading in agarwood. According to the statistics of Chinese Customs, the amount of imports in each province far exceeds that of exports, with Guangdong and Fujian provinces being the main offenders. From 2006 to the first half of 2011, there were 211 cases of smuggling discovered by Guangdong Customs, most of which were carried out by people traveling, transportation, shipping, and express delivery. Because of united efforts by the State Forestry Administration and General Customs Administration of China, the illegal logging and trading has been effectively under control in the past 3 years.

2.6 Concluding Remarks

In China, plantation resources of *Aquilaria* experienced rapid growth recently, to protect and supplement the global wild resources and to supply raw materials to the mounting agarwood industry. To ensure quality supervision of agarwood market, several identification methods, i.e., wood anatomy, DNA barcoding, and chemical analysis, have been demonstrated. However, a reliable international standard containing scientific and rapid-testing technologies to tackle issues in agarwood identification, grading, and quality assurance is still not in place. Efforts should be made at the international level where a practical solution can be suggested and adopted by all concerned parties in the agarwood trade. Only by international coordination, a systematic quality assurance scheme could be inaugurated for agarwood.

References

Cheng J, Yang J, Liu P. Atlas of Chinese woods. Beijing: Chinese Forestry Publishing House; 1992.

Editorial Board of Flora of China of Chinese Academy of Sciences. Flora of China, vol. 52. Beijing: Science Press; 1999.

Endangered Species Import and Export Management Office of the People's Republic of China. Survey report of protection and trade of Chinese agarwood, Beijing, China; 2011.

Eurlings MCM, Gravendeel B. *Trn*L-*trn*F sequence data imply paraphyly of *Aquilaria* and *Gyrinops* (Thymelaeaceae) and provide new perspectives for agarwood identification. Pl Syst Evol. 2005;254:1–12.

Gonzalez MA, Baraloto C, Engel J, Mori SA, Pétronelli P, Riéra B, Roger A, Thébaud C, Chave J. Identification of Amazonian trees with DNA barcodes. PLoS One. 2009;4, e7483.

Heuveling van Beek H, Phillips D. Agarwood: trade and CITES implementation in Southeast Asia. Unpublished report prepared for TRAFFIC Southeast Asia, Malaysia; 1999.

IAWA Committee. IAWA list of microscopic features for hardwood identification. IAWA Bull. 1989;10:219–332.

Ishihara M, Masatsugu Y, Uneyama K. Preparation of (−)-guaia-1(10),11-dien-15,2-olide and (−)-2-hydroxyguaia-1(10),11-dien-15-oic acid, fragrant sesquiterpenes in agarwood (*Aquilaria agallocha* Roxb.). Tetrahedron. 1992;48:10265–76.

Jiao L, Yin Y, Xiao F, Sun Q, Song K, Jiang X. Comparative analysis of two DNA extraction protocols from fresh and dried wood of *Cunninghamia lanceolata* (Taxodiaceae). IAWA J. 2012;33:441–56.

Jiao L, Yin Y, Cheng Y, Jiang X. DNA barcoding for identification of the endangered species *Aquilaria sinensis*: comparison of data from heated or aged wood samples. Holzforschung. 2014;68:487–94.

Konishi T, Konoshima T, Shimada Y, Kiyosawa S. Six new 2-(2-phenylethyl) chromones from agarwood. Chem Pharm Bull (Tokyo). 2002;50:419–22.

Kress WJ, Wurdack KJ, Zimmer EA, Weigt LA, Janzen DH. Use of DNA barcodes to identify flowering plants. Proc Natl Acad Sci U S A. 2005;102:8369–74.

Lin L, Qi S. Triterpenoid from Chinese eaglewood (*Aquilaria sinensis*). Chinese Herbal Med. 2000;31:658–9.

Liu P, Yang J, Lu H. Tropical woods imported from Southeast Asia (version 2). Beijing: Chinese Forestry Publishing House; 2008.

Mei Q, Li H, Wang K, Wu H, Li H. History and origin of "Nanyao" agarwood. Pharmacy Today. 2011;21:3–5.

Miles A. Photomicrographs of the world wood. Building Research Establishment report, Department of the Environment, London. Her Majesty's Stationery Office; 1978.

Morita K. The book of incense. Tokyo: Kodansha International; 1992.

Rachmayanti Y, Leinemann L, Gailing O, Finkeldey R. DNA from processed and unprocessed wood: factors influencing the isolation success. Forensic Sci Int Genet. 2009;3:185–92.

Schlumbaum A, Tensen M, Jaenicke-Despres V. Ancient plant DNA in archaeobotany. Veget Hist Archaeobot. 2008;17:23–244.

The State Council of the People's Republic of China. The first part directory of national protected flora on emphasis. Beijing, China; 1999.

Wyn LT, Anak NA. Wood for trees: a review of the agarwood (gaharu) trade in Malaysia. Petaling Jaya/Selangor/Malaysia: TRAFFIC Southeast Asia; 2010.

Yang Q. The research of chemical composition of agarwood. Nat Product Res Development. 1998;10:99–103.

Chapter 3
Understanding Agarwood Formation and Its Challenges

Saiema Rasool and Rozi Mohamed

Abstract The resinous portion of the *Aquilaria* tree is called agarwood, a valuable non-timber product being used as medicine and incenses in Asia, Middle East, and Europe. Driven by high demand, the wild resources of agarwood-producing trees have been greatly threatened. This fragrant product contains many aromatic substances and is obtained from the pathological conditions of the wood of living trees. The knowledge regarding the technology for inducing agarwood and its continuous formation in the tree is still limited. To conserve the wild *Aquilaria* spp. and to supply sustainable amount of agarwood, cultivation of *Aquilaria* trees in combination with induction through artificial technique is seen as the best approach. In this chapter we will discuss the fundamentals of agarwood formation in the producing trees, the molecular pathway in its synthesis, current methods applied for agarwood induction in cultivated trees, and finally the factors influencing agarwood yield and quality.

3.1 Introduction

Agarwood is a precious non-timber product of tropical tree origin. Due to its aromatic fragrance, it is used as a raw material in making incenses and perfumes. Agarwood is held with high regard in many different cultures of the world, thus the various names: *agar* (Hindi), *agaru* (Tibetans), *akil* (Tamil), *chenxiang* (Chinese), eaglewood (Papua New Guinea), *gaharu* (Malay), *jinkoh* (Japanese), *oud* (Arabic), *mai ketsana* (Laos), *mai kritsana* (Thai), *sasi* or *sashi* (Assamese), and *tramhuong* (Vietnamese). The Japanese *jinkoh* literally means "sinking incense," and the highest grade is known as *kyara* (Kiyoko 1992). In Europe, agarwood is historically known as *Lignum aquila* (eaglewood)/agilawood, *Lignum aloes*, or aloeswood (Henry and Burnell 1903).

Agarwood has high commercial value but the tree source is severely endangered due to indiscriminate felling. Over the years, demand for this scarce material,

S. Rasool • R. Mohamed (✉)
Department of Forest Management, Faculty of Forestry, Universiti Putra Malaysia, UPM, Serdang, Selangor 43400, Malaysia
e-mail: rozimohd@upm.edu.my

R. Mohamed (ed.), *Agarwood*, Tropical Forestry,
DOI 10.1007/978-981-10-0833-7_3

which is often obtained from natural population, has increased tremendously. This has led to its exploitation in sourcing countries such as India, China, and Southeast Asian countries, together with West Papua and Papua New Guinea (PNG) (Barden et al. 2000; TRAFFIC Oceania, 2002; TRAFFIC East Asia-Taipei and TRAFFIC Southeast Asia, 2005; Wyn and Anak 2010). Agarwood is exported in the form of blocks, wood chips, powder, dust, and essential oil. All these come in a wide range of grades and thus prices. To overcome demand and conserve the sources, much effort has been drawn toward cultivating the tree species in sourcing countries, such as Cambodia, China, Indonesia, Malaysia, Thailand, and Vietnam.

3.2 Agarwood-Producing Tree Genera

The Thymelaeaceae family consists of many important incense-producing tree species, including those from the genera *Aquilaria*, *Gyrinops*, *Gonystylus*, and *Aetoxylon*. While all have the ability to produce incense wood, *Aquilaria* and *Gyrinops* are the two most important genera when it comes to producing agarwood. *Gonystylus* is better known for its hardwood timber from the peat swamp, whereas *Aetoxylon* produces fragrant wood of less importance, by comparison.

The genus *Aquilaria* has a wide distribution in the Indomalesian region. This is reflected from the wealth of names given to this highly valuable wood in different cultures. For a long period of time, *Aquilaria* has been harvested severely from the forests forcing their numbers to decline. The species *A. malaccensis*, in particular, was so heavily poached that in 1995, it became the first species subjected to the trading regulation of the Convention of the International Trade of Endangered Species (CITES). Worldwide, there are 21 *Aquilaria* species and nine *Gyrinops*; these have been reviewed and accepted (The Plant List, version 1.1). Many of them have significant profitable value. Among the *Aquilaria* genus that have been exploited are *A. beccariana*, *A. crassna*, *A. filaria*, *A. malaccensis*, *A. microcarpa*, *A. sinensis* and *A. subintegra*, while from the *Gyrinops* are *G. ledermannii*, *G. versteegii*, and *G. walla*. Because of their close resemblance, it was difficult to ascertain species identity. This effectively placed all *Aquilaria* spp. and *Gyrinops* spp. under CITES listing in 2005. *Gonystylus* was not spared as well but for reasons associated to timber logging. The listing under Appendix II of CITES is perceived as an international driving force toward a legal and sustainable manner of agarwood trading between importing and exporting countries (Compton 2004). CITES is one of the world's commanding agreements on species conservation. Its method of controlling is by employing trade permit and by imposing a quota on the export quantity by country members. The rationale of doing this is to protect agarwood from overharvesting, regardless of species.

Aquilaria has been studied in more detail when compared to *Gyrinops*; thus it can be considered as the model tree for incense-producing species. In this chapter, special attention is given to *Aquilaria* due to the current mass of scientific knowledge gathered on this genus.

3.3 The Beginning of Agarwood in the Wood Tissue

Agarwood is an oleoresin; it has a solid or semisolid form (viscous) and it is insoluble in water. Like many softwood tree species, the agarwood resin is secreted in the stem tissues of *Aquilaria*. However, in *Aquilaria*, bundles of phloem cells are produced throughout the xylem as well as in a layer external to the xylem. This is in contrast to angiosperms that produce phloem cells growing out from the circumference of the cambium. Indeed, wood anatomy of *Aquilaria* is exceptional because the xylem region, in addition to vessels, fibers, and parenchyma cells, also consists of groups of phloem cells called included phloem or interxylary phloem. Living parenchyma cells are the most important elements in agarwood formation because they are able to biosynthesize resinous substances of agarwood (Nobuchi and Mohd Hamami 2008). Brownish droplets containing the resinous substances have been detected in the included phloems, parenchyma cells, and vessels, in the xylem region of agarwood-containing stem (Fig. 3.1) (Mohamed et al. 2013). This network of anatomical structures appears to be responsible in producing, storing, and distributing agarwood constituents to affected areas as the tree defends itself against various enemies and damages. Previously, it was thought only old trees produce resinous agarwood. New evidence has emerged that even juvenile *Aquilaria* trees own these structures and are capable to produce agarwood when induced (Mohamed et al. 2014b).

To understand agarwood formation in *Aquilaria*, it is valuable to look into another tree family that produces resin, the Pinaceae. In pine and spruce trees, interconnecting vertical and horizontal resin canals are regular features of the wood. The canals are tubular structures composed of long-lived epithelial resin-secreting cells,

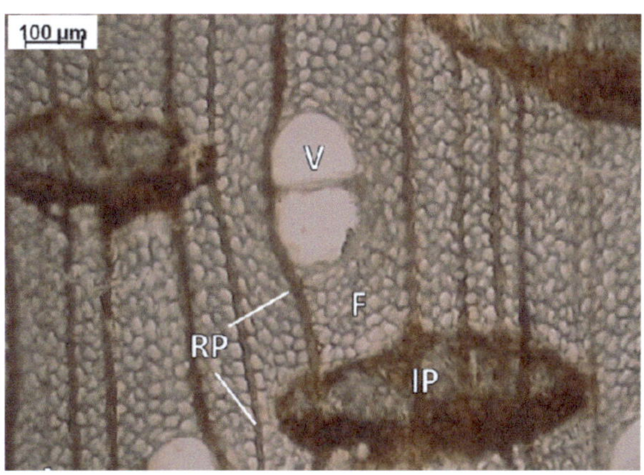

Fig. 3.1 A transverse section of an agarwood-containing stem. Brownish substances are seen deposited in the included phloems (*IP*) and ray parenchyma (*RP*) cells in this unstained section (Scale bar = 100 µm). *V* vessel, *F* fiber (Source: Mohamed et al. 2013)

parenchyma support cells, and strand tracheids. Resin is formed in the epithelial cells and is transported through the matrix of vertical and horizontal ducts (reviewed in Cown et al. 2011). Heartwood formation (resin impregnation) commences when the stems progresses outward from the pith, and this is most apparent in old pine stems (25–30 years) where the resin may account for 25–35 % of the weight of the heartwood (reviewed in Cown et al. 2011). In relation to *Aquilaria*, the oleoresin filled up the various compartments, became hard, and got impregnated in the wood. The extensive deposits of resin over time increase the wood density of agarwood to unusually high levels.

3.4 Causes to Agarwood Formation

General observations established that the fragrant resin is never produced in sound trees. This suggests an association exists between damaged tissues and infestation by biotic agents, which leads to the resulting agarwood (Oldfield et al. 1998). Indeed, wounding has been recognized as the first cause to agarwood formation (Pojanagaroon and Kaewrak 2005; Nobuchi and Siripatanadilok 2008). Physical damage causes the tree to weaken and become vulnerable to fungal infection. Soon after, in a matter of days, a slight discoloration would become visible surrounding the wound. It darkens with time and is easily detected against the otherwise white-wood. The darkened zone eventually gets bigger and signifies agarwood presence. This proves that agarwood formation may involve prolonged microbial infection, which continuously elicits synthesis of agarwood constituents in the area. Damage by boring insects, for example, involves wounding followed by infection. Fungi, in particular, are believed to be the major infecting agent that enters the host through the wounds. The host's defense system reacts by producing agarwood compounds to fight off the pathogens. Since the 1900s, many workers have associated the agarwood zones to fungal infection (reviewed in Ng et al. 1997). Direct fungal isolation, microscopy, and more recently rDNA cloning and sequencing have revealed that the area in and surrounding agarwood zones do harbor a variety of fungi. Among them are members of the genera *Aspergillus*, *Cunninghamella*, *Curvularia*, *Fusarium*, *Lasiodiplodia*, *Penicillium*, *Pythium*, *Trichoderma* (Mohamed et al. 2010; Premalatha and Kalra 2013), and many others.

In its natural surroundings, *Aquilaria* is highly exposed to a diverse group of microorganisms, whether in the rhizosphere in the soil (Nimnoi et al. 2011) or in the stems itself (Zhang et al. 2014b). There are limited studies regarding *Aquilaria*'s interaction with its diverse fungal community. In one study, three fungal species in two wounded wild *A. malaccensis* were tracked over time using quantitative real-time PCR (qPCR), in wounds that represent different time points, i.e., 0–18 h, 2–13 days, 2–18 weeks, and 6–12 months (Mohamed et al. 2014a). The three species were common fungi often associated with agarwood formation: *Cunninghamella bainieri*, *Fusarium solani*, and *Lasiodiplodia theobromae*. The qPCR data revealed that the abundance of the three species decreased over time. The fungi were detected

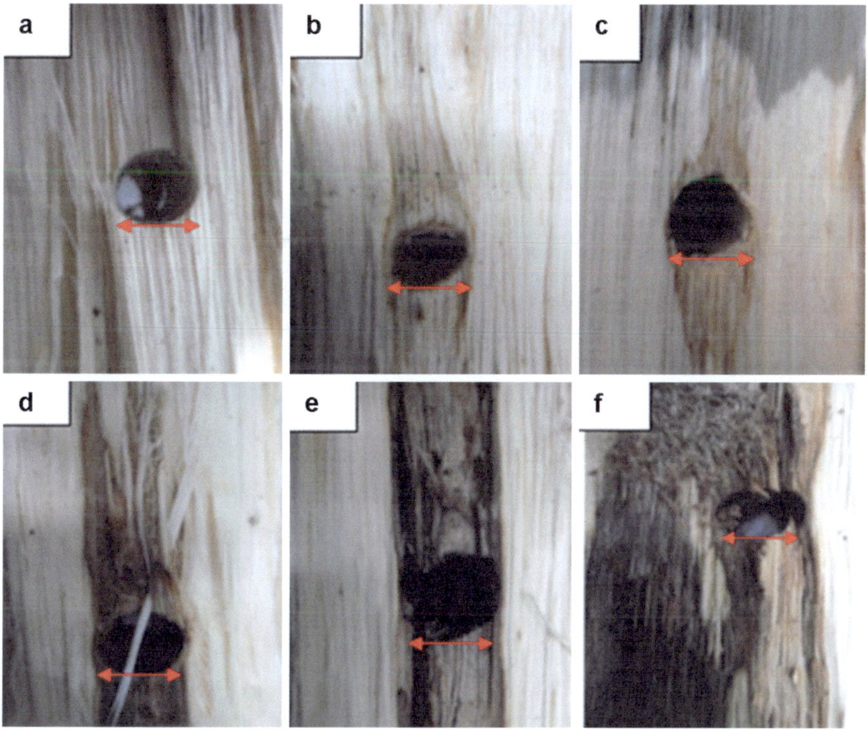

Fig. 3.2 The progression of the darkened zones observed on wounded stem surfaces at different times after wounding. (**a**) 2 days, *light yellow*; (**b**) 5 days, *light brown*; (**c**) 2 weeks, *brown*; (**d**) 4 weeks, *dark brown*; (**e**) 6 months, *blackish brown*; and (**f**) 9 months, *black*. The diameter of the hole is 16 mm (*red arrow*) (Source: Mohamed et al. 2014a)

in high numbers during the first few hours and days after wounding and in low numbers many months later. On the contrary, the darkened zone where agarwood is formed continued to enlarge over time (Fig. 3.2). The accumulation of agarwood compounds at the wounding site could have caused the decline in fungal abundance, which is consistent with its role in defense response.

3.4.1 Customary Induction

Agarwood induction, as perceived by local people, on trees growing in the natural environment, happened in old trees that had been stroked by lightning or attacked by animals, insects, or microbes. The formation is usually in proximity to wounded or decaying parts of the trunk. Initially it was thought that formation of agarwood takes place only in the stem or the main branches. New observations however revealed that it happens in roots and twigs as well. The production of agarwood was thought to happen very slowly in old trees (Gianno 1986; Chakrabarty et al. 1994), and the slow process due to fungal involvement in infecting the wood. New studies revealed that even at the age of 3 years, cultivated trees can produce agarwood when artificially treated (Heuveling van Beek and Phillips 1999; Mohamed et al. 2014b).

Various conventional methods have been developed. For example, the Chinese farmers have been using the burning-chisel-drilling and partial-trunk-pruning methods. Other methods of wounding by using ax, machete, bark removal, and nailing have also been useful in other parts of the world such as in Indonesia, Malaysia, and Thailand. These methods are hand-down traditions and have been practiced sustainably on trees growing on individual and ancestral lands and in natural forests.

3.4.2 Non-customary Induction

Over decades, people pondered on how agarwood forms in nature. From the information gathered, nonconventional methods had emerged to mimic the natural event. Many researchers began to employ artificial induction by inoculating inducing agents directly into the tree stem. Early works have been documented by Ng et al. (1997), beginning with Tunstall in 1929, who first inoculated fungi experimentally into *Aquilaria agallocha* followed by others such as Bose (1934), Sadgopal and Varma (1952), Gibson (1977), Rahman and Khisa (1984), and Santoso (2013). All applied open wounds as the mode of inoculation with the only difference being the fungal species. The outcomes of their inoculation tests were of mixed results. At that point, it was concluded that agarwood formation was initiated by wounding followed by nonspecific fungal infection (Rahman and Basak 1980).

3.4.3 Modern Artificial Induction

To meet the demand and market needs of agarwood, large plantations have been established in sourcing countries. This trend has been increasing for the last 20 years with participation from individuals, local communities, entrepreneurs, and government agencies. Although planting trees is a straightforward venture, agarwood induction is quite complicated. Without an efficient and proven induction method, return on investment made in the plantations will be minimal because the healthy wood is of little value. To develop proven methods, two directions have been considered: (1) the delivery mode and (2) superior inducing agents, whether in the form of microbes, chemicals, or both. One such method was pioneered by Blanchette and Heuveling van Beek (2009) known as Cultivated Agarwood kits (CA-Kits), where tubes are placed in the tree trunk as a mean to introduce microbes and to arouse production of the defense compounds by the tree naturally. Others such as the Taiwan and Pheerapan methods are also available in the market. Both methods use microbes for the production of agarwood (Chang et al. 2011); however, it is a slow process.

3.4.3.1 Inoculum-Based Inducer

Generally, inoculum- or microbe-based techniques require a long incubation time before harvesting to produce darker wood and subsequently better forms of agarwood. In a study to evaluate the effect of several fungal species on agarwood formation over time in *A. malaccensis*, changes in the length and intensity of the darkened zone were observed after 3- and 6-month periods following inoculation (Mohamed et al. 2014b). A positive relationship with time was perceived but not with the tested fungal species. Gas chromatography-mass spectrometry (GC-MS) analysis of the 6-month-old sample yielded some important agarwood compounds such as benzylacetone, anisylacetone, guaiene, and palustrol. Similar results were obtained from artificially inoculated *A. sinensis* with the fungus *Melanotus flavolivens* when samples were harvested 6 months and 1 year post inoculation (Lin et al. 2010). Major agarwood compounds – specifically benzaldehyde, benzenepropanoic acid, anisylacetone, and a chromone, 8-methoxy-2-(2-phenylethyl)-4H-1-benzopyran-4-one – were detected. Other successful inocula that have been tested include the ascomycete fungus *Paraconiothyrium variabile* (Cui et al. 2013) and the deuteromycete *Lasiodiplodia theobromae* (Zhang et al. 2014a) on *A. sinensis* and *Fusarium oxysporum* and *Fusarium solani* on *A. microcarpa* (Akhsan et al. 2015). However, less promising results for artificial fungal induction methods were also reported (Tamuli et al. 2005; Bhuiyan et al. 2009). When under attack, the first defense mechanism activated by *Aquilaria* is observed in the form of callus around the wounded area where agents gained penetration into the host's cell (Blanchette and Heuveling van Beek 2009). It can be speculated that *Aquilaria* choose to perform callusing rather than producing resin when the threat level is weak and nonpersistent. In view of this, to develop an effective inoculum-based inducer, interaction between inoculum strain and *Aquilaria* genotypes needs to be explored further.

3.4.3.2 Chemical as Inducer

Chemicals appear to be potent agarwood inducers. Wei et al. (2010) showed that chemicals when coupled with an ingenious mode of delivery system can induce agarwood in the whole tree and not confined to local areas as shown in other conventional methods. This promising method, patented as "whole-tree agarwood-inducing technique" (Agar-Wit), applies simple and cheap transfusion sets through which agarwood inducers are injected into the xylem part of the tree (Zhang et al. 2012). The inducer is in liquid form and is elated to the whole body of the tree due to water transportation causing agarwood to form in every woody part. Most importantly, the induced agarwood meets the medicinal quality set by the Chinese Pharmacopoeia (Liu et al. 2013). The identities of chemicals in patents are generally not made public, or their strengths and formulations are kept discrete. They could be in the form of plant defense elicitors such as hydrogen peroxide (H_2O_2), methyl jasmonate (MJ), or salicylic acid (SA) (Okudera and Ito 2009; Wijitphan 2009; Wei et al. 2010) or chemicals such as sodium chloride, sulfuric acid, formic acid, and sodium methyl bisulfite (Thanh et al. 2015). Chemical-based inducer acts by causing severe injury to *Aquilaria* cell structure, and as a result callusing cannot be performed to cover up the wound. While some chemicals such as organic based may be safe to consume, there are some unscrupulous individuals who applied pesticides as type of inducer. The impact of this inducer is a great concern to the environment and agarwood consumers. Chemical-based inducers appeared to be very attractive due to the ease of preparation and application, inexpensive, and fast result, but more evidence is needed to demonstrate the effectiveness. More importantly, appropriate tests must be conducted to establish the chemical residual content in the resulting agarwood that does not become a health hazard.

3.5 Compounds in Agarwood

The main compounds found in the oleoresin of agarwood are a complex mixture of sesquiterpenes and 2-(2-phenylethyl) chromones (reviewed in Naef 2011). Together with some simple volatile aromatic compounds, they create this impressive pleasing odor that some described as balsamic, spicy, woody, and sweet. Analysis into the chemical constituents of agarwood had actively begun in the 1960s and is continuing into the twenty-first century. Thus far, the total number of identified compounds is over 150. This explains the richness and diverse organoleptic properties found in agarwood from various species and regions. Agarwood is rich in terpenoid contents, but as of today none of these terpenoids can be manufactured industrially, thus the reliance on natural sources. Attempts have been made to categorize *Aquilaria* and *Gyrinops* species according to their chemical profiles. However, this seemed unattainable because the composition is highly dependent on geographical sites.

Sesquiterpenes in agarwood were first characterized from *A. agallocha* by Indian chemists, more than 40 years ago (reviewed in Konishi et al. 2002). Some 70 sesquiterpene compounds have been identified so far and their structures elucidated (Naef 2011). Examples are agarofurans, eudesmanes, and guaienes and their oxidized forms such as jinkoh-eremol and agarospirol. Both of the latter compounds are known to have sedative and analgesic effects (Okugawa et al. 1996; Takemoto et al. 2008). Another characteristic for agarwood is chromone. The first report on oxygenated chromone derivative from agarwood was released in 1978 (Yoshii et al. 1978). Since then, many more structurally different chromones have been discovered. Close to 40 2-(2-phenylethyl) chromones have been recognized from agarwood of numerous abilities, of which 17 are agarwood specific and may be used as phytochemical markers for authentication purposes (reviewed in Naef 2011). In addition, three derivatives of diepoxy-tetrahydro-2-(2-phenylethyl) chromones may be used to indicate infected wood because they appeared only in resins isolated from wounded wood of *Aquilaria* and not in the healthy part (Yagura et al. 2005). Interestingly, the trio also appeared in *Aquilaria* calli and cell cultures in addition to guaiene-derived compounds (Okudera and Ito 2009) indicating that in vitro agarwood production is possible. Chromone compounds are described as having bigger role in producing the warm, balsamic, and enduring odor when agarwood is heated and smoke is produced. In very high-quality agarwood such as *kanankoh*, the chromone content is reportedly at 60 %, while the less superior type *jinkoh* has only 1.5 % (Ishihara et al. 1993). *Kanankoh* is used in the *kohdoh* ceremony (listening to incense) where the fragrance of agarwood is appreciated in an elaborated traditional manner.

The types and derivatives of sesquiterpenoids and chromones in agarwood are extensive. After the comprehensive review by Naef (2011), additional new compounds have been discovered from agarwood (Wu et al. 2012; Li et al. 2014, 2015; Yang et al. 2014; Wang et al. 2015), and the number no doubt will continue to grow.

3.6 Biosynthesis of Major Compounds

From general understanding, it can be concluded that the most probable role of agarwood is for the tree to defend itself against biotic and abiotic stresses. Stress induces the defense response and triggers the secondary metabolism network leading to agarwood compound formation and resin accumulation. Essentially, agarwood gets its fragrance from the presence of the aromatic terpenes, specifically the sesquiterpenes, and the chromones (Naef 2011). For that reason it is important to understand the synthetic mechanisms of these two compounds. In this section, special emphasis is given to the activation of genes in the sesquiterpenoid synthetic pathway.

3.6.1 Sesquiterpenoid Biosynthetic Pathway

Much information has been gathered on terpenoid biosynthetic pathways (Hu and Lu 2015; Singh and Sharma 2015). In plants, two pathways have been established: (1) the mevalonic acid (MVA) and (2) the 1-deoxy-D-xylulose-5-phosphate (DXP), also known as the methylerythritol phosphate (MEP) pathways. They manufacture the C5 isoprene units, isopentenyl diphosphate (IPP), and its isomers, dimethylallyl diphosphate (DMAPP). These are important terpenoid building blocks, being synthesized either in the cytoplasm or the plastid organelle, according to the respective pathway. The formation of IPP and DMAPP from acetyl-CoA or pyruvate are catalyzed by a sequence of different enzymes. Genes encoding for these enzymes have been identified and characterized from *Aquilaria* species via transcriptome (Xu et al. 2013) and genome sequencing (Chen et al. 2014) and are discussed in more detail in Chap. 5. In the following step, IPP or DMAPP is connected through the head-to-tail connections to form farnesyl diphosphate (FPP) (C15 unit), the precursor of sesquiterpenes, in the presence of FPP synthase (FaPS) (Cao et al. 2012). Kenmotsu et al. (2011) first cloned the *Am-FaPS-1* gene from *A. microcarpa*, which exhibited high homology with FaPS from different plant sources. The transcript was abundant upon exposure of the cell culture to MJ, yeast extract, and Ca^{2+} ionophore A23187, indicating that the two former substances are triggers to induced responses in plants, while Ca^{2+} acts as a molecule messenger in activating the process. This clearly shows that substances such as MJ could be used to enhance biosynthetic pathways of secondary metabolites in *Aquilaria*.

In the final step, the FPPs are transformed into sesquiterpenes (C15) by specific enzymes; sesquiterpene synthases. Genes encoding for sesquiterpene synthases in *Aquilaria* are present in multiple copies (Kumeta and Ito 2010). At least five clones have been reported from *A. crassna*, all having high similarities in amino acid sequences. After being expressed in *E. coli* and the product enzymatically assayed using FPP, only three clones generated the same compounds as mined from MJ-treated cells. The product was δ-guaiene. These genes and their encoded enzymes are the first sesquiterpene synthases yielding guaiene-type sesquiterpenes as the major products. Via transcriptome sequencing, Xu et al. (2013) have identified several clones of sesquiterpene synthases (*ASS1*, *ASS2*, and *ASS3*) from *A. sinensis*, all yielding guaiene-type product as well. Despite its richness in sesquiterpenes, genes encoding for sesquiterpene synthases producing other types of sesquiterpenes have not been reported from *Aquilaria*.

3.6.2 Signaling Pathway

Wounding, insect boring, and fungal infection are types of abiotic and biotic stresses that can provoke agarwood production in *Aquilaria*. *Aquilaria* reacts to wounding and pathogen attack by stimulating specific genes of which some are expressed in the area of wound site, and the rest is triggered in the non-damage part of the plant, via the activation of multiple signaling events, similar to other plants' response to stress (Mucciarelli et al. 2007; Rodriguez et al. 2009; Wang et al. 2010). One pathway that has been proposed for *Aquilaria* is the MAPK (mitogen-activated protein kinase) signaling pathway (Xu et al. 2013). Throughout the eukaryotic evolution, MAPK cascades have been highly conserved modules (Kusari et al. 2004; Pitzschke et al. 2009). These cascades are nominally made up of MAPK kinase kinase (MAPKKK), a MAPKK (MAPK kinase), while MAPK connects the upstream receptors to downstream targets. Using *A. sinensis* as the model tree, it has been proposed that at transcriptional level, the *ASS1* gene expression is controlled by wound signal, which activates the MAPK cascade and phosphorylates downstream transcription factors (TFs) like MYB or WRKY (Xu et al. 2013). The initiation of *ASS1* transcription must be due to triggered TFs that bind to the cis-acting elements in the promoter of *ASS1*. About 41 unigenes were interpreted to be correlated to the

MAPK pathway and another 25 to calcium signaling that may perform roles in the wound responses of agarwood formation in *A. sinensis*. The treatment of MJ showed considerable upregulation of transcription factors like MYB4, WRKY4, MAPK2, MAPKKK, and the NADPH oxidase. Elicitors like jasmonic acid (JA) are known to trigger the TFs downstream through the hydrogen peroxide pathway (Kazan and Manners 2008). Some of the TFs with positive regulators of sesquiterpene synthases like the *AP2*, *WRKY*, and *MYC* genes have been identified from *A. sinensis* (Xu et al. 2013) and their functions compared to their homologues in other plant species. In *Gossypium arboreum*, the *GaWRKY1* transcription factor shows positive expression of (+)-δ-cadinene synthase that catalyzes the biosynthesis of sesquiterpene gossypol (Xu et al. 2004). In *Artemisia annua*, the *ADS* gene regulates the biosynthesis of artemisinin in the presence of the TF, *AaWRKY1* (Ma et al. 2009). The responsive *AP2/ERF* (ethylene-responsive factor) was also found to control the biosynthesis of artemisinin by binding to *CBF2* and *RAA* (Yu et al. 2012). Similarly, when *Oryza sativa* was treated with MJ, a marked increase in the expression of the sesquiterpene synthase gene, *TPS3*, was observed, in addition to the release of more than ten sesquiterpenes mainly of the β-caryophyllene type (Cheng et al. 2007).

3.6.3 Chromones

Chromones are one of the major classes of naturally occurring compounds including flavonoids and possess important biological activities as antitumor, antioxidant, anti-inflammatory, antibacterial, and many more (Tawfik et al. 2014). Chemically, chromones (4H-chromen-4-ones) are heterocyclic compounds with the benzopyranone ring. *Aquilaria* spp. is one of the few plant species that produces the rare chromone known as 2-(2-phenylethyl) chromone. To our knowledge, there is no report on the biosynthesis of chromones in *Aquilaria*, although generally they are thought to be a result from convergence of multiple biosynthetic pathways such as the acetate, pentaketide, and shikimate pathways. This chromone group has been discovered only in a handful plant species (reviewed in Ibrahim and Mohamed 2015). Due to the vast range of biological functions associated with this compound scaffold, several synthetic applications have been developed to find new chemical entities as new drugs using the chromone ring system as the backbone structure. Advances in chemical processes have shown that the synthesis of 2-(2-phenylethyl) chromone is possible (Goel et al. 2006).

3.7 Factors Influencing Agarwood Yield and Quality

The yield and quality of the resinous agarwood vary considerably. Gianno (1986) suggested that a tree above 20 cm in diameter at breast height produces approximately 1 kg of agarwood. However, research in West Kalimantan, Indonesia, shows that the yield of *Aquilaria* resin is not correlated with tree diameter or timber volume, even when the trees have similar progression in infection (Soehartono and Mardiastuti 1997). Several factors are thought to affect the produced agarwood, both in quantity and quality (Ng et al. 1997). In our opinion, the most important factor is genetic variability between the tree species themselves. Unfortunately, no information is available regarding genetic variation or heritability for *Aquilaria* or *Gyrinops*. This is partly due to the difficulty in quantifying and ascertaining quality of the produced agarwood. Another important factor is the treatment applied onto the tree to induce agarwood. In a field trial conducted in planted *A. crassna*, Thanh et al. (2015) showed that agarwood oil samples from different treatments including biological (fungal mixture), chemical (sulfuric acid and sodium methyl bisulfite mixture), and mechanical (hammered nails) differ in quantity and quality. A year after, chemical treatment yielded agarwood oil with the highest sesquiterpene content compared to the other treatments and non-treated. On the other hand, biological treatment gave the highest sesquiterpene yield when the trees were left for 2 years. Time appears to be a major factor when biological agent is involved.

To grow *Aquilaria* species or genotype in its native climatic region is perhaps more suitable compared to foreign species. However, it has been shown that *Aquilaria* adapts easily to new environment, resulting in many *Aquilaria* plantations in sourcing countries been planted with nonnative species. For example, successful *Aquilaria* plantations have been established in Australia, a region that is beyond the tree natural distribution (Page and Awarau 2012). In Malaysia, where *A. malaccensis* is native, people have succeeded in planting *A. crassna* and *A. subintegra* species as well. One problem that could arise from this situation is the complex relationship between inoculum, tree genotype, and the environment. The type of inoculum or inducer is an important factor that influences agarwood quality. Each strain of inoculum certainly propagates better in its own environment; therefore a broad-spectrum inoculum is needed for economic benefits. Similar strains when applied on *Aquilaria* and *Gyrinops* trees yielded agarwood of differing smells (Turjaman and Santoso 2012). If the host factor is removed, and the inoculum is replaced by chemical inducer, the resulting agarwood could be of a more consistent quality (Liu et al. 2013). This proves that agarwood quality is controlled by genetic factors of the host and the inoculum, with the environment playing a compounding role.

Agarwood is in the trade history since long time ago, but its quality is very subjective and highly dependent on personal experiences. In the past, collectors harvested agarwood from wild trees, and quality determined from the age of the tree. It was thought that an older tree yielded higher agarwood quality (Barden et al. 2000; Persoon 2007). In contrast to this common belief, 7 to 8 years old trees are found capable of producing agarwood (Paoli et al. 2001; Chetpattananondh 2012). Even trees as young as 3 to 5 years old have been shown to form agarwood (Xu et al. 2013; Gao et al. 2014; Mohamed et al. 2014b). Therefore, it is not the age of the tree but the use of proper treatment that is paramount for inducing agarwood. It has now become evident that plantation trees can yield agarwood with quality similar to high-grade wild agarwood when using suitable induction methods such as the Agar-Wit (Liu et al. 2013). More on the different agarwood grades and grading methods are discussed in Chap. 10.

3.8 Valuable Agarwood Compounds from Cell Suspension Culture

Biotechnology offers an opportunity to exploit new means for the production and accumulation of many of the valuable chemical compounds found in plants such as alkaloids, terpenoids, steroids, saponins, phenolics, flavanoids, and amino acids, through plant cell cultures. Even though there are limitations in using plant cell cultures, some can produce higher amount of secondary metabolites than the intact plants (Sree et al. 2010). At present, only limited work on plant cell culture of *Aquilaria* has been reported. The production of valuable agarwood compounds has been shown feasible in cell suspension culture of *Aquilaria* when induced with a proper elicitor. Three species of sesquiterpene (á-guaiene, á-humulene, and δ-guaiene) and four of chromones (phenylethylchromones (5S,6R,7R,8S)-2-(2-phenylethyl)-5e',6e,7a,8a'tetrahydroxy-5,6,7,8- tetrahydrochromone; 6-hydroxy-2-(2phenylethyl) chromone; 6-methoxy-2-(2-phenylethyl) chromone; 6-methoxy-2-[2-(3-methoxyphenyl)ethyl] chromone; 6,7-dimethoxy-2-(2-phenylethyl) chromone) were found to be induced by molecules in signaling transduction such as MJ and salicylic acid (Ito et al. 2005; Okudera and Ito 2009). Elicitors are signal molecules responsible for triggering the signal transduction cascade leading to the activation and expression of genes in the biosynthesis of secondary metabolites (Wang and Wu 2013). In *Aquilaria*, induction can be in the form of abiotic or biotic factors; as such, elicitors could be molecules that are excreted by the pathogen or the fungal propagules themselves. Cell suspension culture of *A. sinensis* produced four derivatives of 2-(2-phenylethyl) chromones when challenged with crude fungal extracts of *M. flavolivens* (Qi et al. 2005). In another study (Jayaraman and Mohamed 2014), crude mycelial extracts of *Trichoderma* sp. elicited several important agarwood compounds including 8-epi-.gamma.-eudesmol, á-guaiene, and alloaromadendrene oxide-1. The elicitor was added to the cell suspension culture, initiated with fresh calli originated from the leaf explants of *A. malaccensis* (Fig. 3.3). These are promising results on the potential use of fungal elicitor as biological inducer for valuable agarwood compound production in in vitro *Aquilaria* cultures.

3.9 Conclusion and Future Perspectives

Since the ever increasing international demand for agarwood cannot be satisfied by limited natural stocks of *Aquilaria* trees, there is a need for mass cultivation of the trees to relieve the pressure on its natural population. Not only planting better genotypes can help, but one has to develop techniques that are efficient at inducing agarwood and practical for application in plantation setting. To achieve this goal, background knowledge of the biology of the organism is important. Therefore, understanding agarwood induction and formation, specifically at molecular level, is essential for improving the production in living trees. By

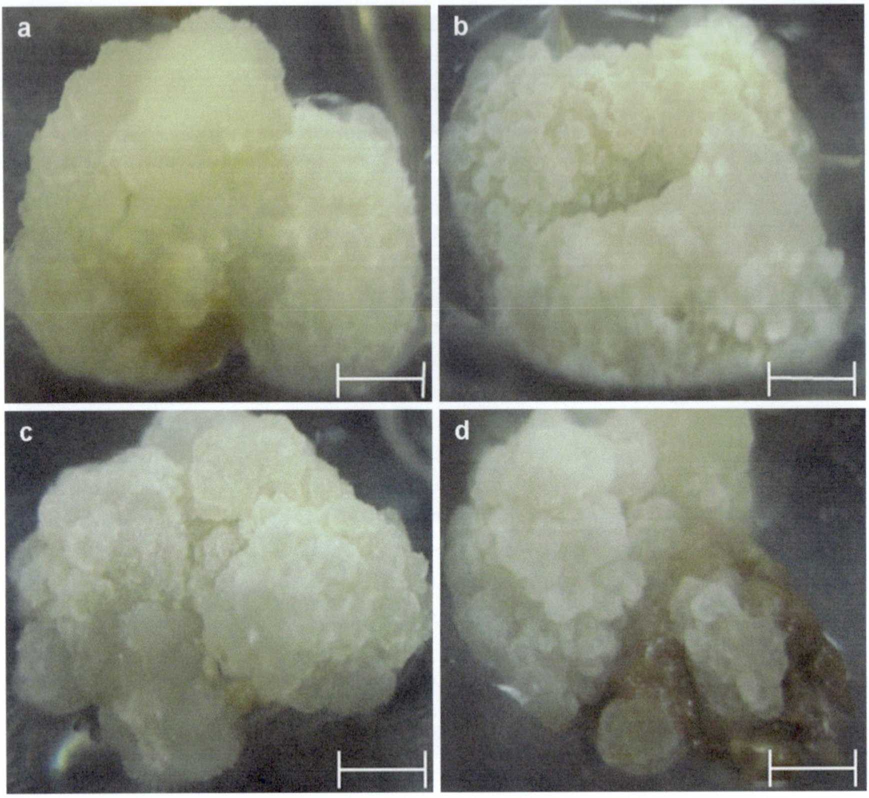

Fig. 3.3 Callus induced from *Aquilaria malaccensis* leaf explants after 30 days incubation in MS medium supplemented with 1.1 µM naphthaleneacetic acid (NAA) and various 6-benzylaminopurine (BAP) concentrations, (**a**) 0.55 µM, (**b**) 1.1 µM, (**c**) 2.2 µM, (**d**) 3.3 µM. Bar = 5 mm (Jayaraman et al. 2014)

elucidating the biosynthetic pathway and regulation of induction, better ways and techniques can be developed to boost the success of agarwood formation. In the future, novel techniques created from this knowledge may be used to supply quality agarwood from plantation trees and thus satisfy not only the demand but most crucially to help to preserve wild *Aquilaria* trees in their already depleting natural populations.

References

Akhsan N, Mardji D, Sutisna M. Response of *Aquilaria microcarpa* to two species of *Fusarium* under two different cultivation systems. J Trop For Sci. 2015;27:447–55.

Barden A, Anak NA, Mulliken T, Song M. Heart of the matter: Agarwood use and trade and CITES implementation for *Aquilaria malaccensis*. Cambridge, UK: TRAFFIC International; 2000.

Bhuiyan MNI, Begum J, Bhuiyan MNH. Analysis of essential oil of eaglewood tree (*Aquilaria agallocha* Roxb.) by gas chromatography mass spectrometry. Bangladesh. J Pharma. 2009;4:24–8.
Blanchette RA, Heuveling van Beek H. Cultivated agarwood. US Patent 7638145 B2; 2009.
Bose SR. The nature of agar formation. Sci Cult. 1934;4(2):89–91.
Cao X, Yin T, Miao Q, Li C, Ju X, Sun Y, Jiang J. Molecular characterization and expression analysis of a gene encoding for farnesyl diphosphate synthase from *Euphorbia pekinensis* Rupr. Mol Biol Rep. 2012;39:1487–92.
Chang YS, Nor Azah MA, Rashid AMA. Inducement of gaharu and potential of gaharu oils. In: Rashid AMA, Zuhaidi YA, editors. Tapping the wealth from karas (*Aquilaria malaccensis*) tree. Malayan Forest Records No. 50, Forest Research Institute Malaysia, Kepong. 2011. p. 48–63.
Chen CH, Kuo TCY, Yang MH, Chien TY, Chu MJ, Huang LC, Chen CY, Lo HF, Jeng ST, Chen LFO. Identification of cucurbitacins and assembly of a draft genome for *Aquilaria agallocha*. BMC Genomics. 2014;15(578):1–11.
Cheng AX, Xiang CY, Li JX, Yang CQ, Hu WL, Wang LJ, Lou YG, Chen XY. The rice (E)-beta-caryophyllene synthase (OsTPS3) accounts for the major inducible volatile sesquiterpenes. Phytochemistry. 2007;68:1632–41.
Chetpattananondh P. Overview of the agarwood oil industry. International Federation of Essential Oils and Aroma Trades (IFEAT) International Conference Singapore; 2012. p. 131–8.
Compton J. Final draft report to FAO TCP/PNG/2901, eaglewood management project. Southeast Asia: TRAFFIC; 2004.
Chakrabarty K, Kumar A, Menon V. Trade in agarwood. New Delhi: TRAFFIC India and WWF-India; 1994.
Cown DJ, Donaldson LA, Downes GM. A review of resin features in radiata pine. New Zea J Forest Sci. 2011;41:41–60.
Cui J, Wang C, Guo S, Yang L, Xiao P, Wang M. Evaluation of fungus-induced agilawood from *Aquilaria sinensis* in China. Symbiosis. 2013;60:37–44.
Gao ZH, Yang Y, Zhang Z, Zhao WT, Meng H, Jin Y, Huang JQ, Xu YH, Zhao LZ, Liu J, Wei JH. Profiling of microRNAs under wound treatment in *Aquilaria sinensis* to identify possible microRNAS involved in agarwood formation. Inter J Biol Sci. 2014;10(5):500–10.
Gibson I. The role of fungi in the origin of oleoresin deposits (Agaru) in the wood of *Aquilaria agallocha*. Roxb Bano Biggyan Patrika. 1977;6(1):16–26.
Gianno R. The exploitation of resinous products in a low land Malayan forest. Wallaceana. 1986;43:3–6.
Goel S, Shashi, Makrandi JK. Synthesis of 2-(2-phenylethyl) chromones. Indian J Chem 2006; 45B:535–6.
Heuveling van Beek H, Phillips D. Agarwood: trade and CITES implementation in Southeast Asia. Unpublished report prepared for TRAFFIC Southeast Asia, Malaysia; 1999.
Henry Y, Burnell AC. "Eaglewood" Hobson-Jobson: a glossary of colloquial Anglo-Indian words and phrases, and of kindred terms, etymological, historical, geographical and discursive. 2nd ed. London: John Murray; 1903. p. 335.
Hu C, Lu W. Insight into yeast: a study model of lipid metabolism and terpenoid biosynthesis. Biotech Appl Biol Chem. 2015;62:323–8.
Ibrahim SRM, Mohamed GA. Natural occurring 2-(2-phenylethyl) chromones, structure elucidation and biological activities. Nat Prod Res. 2015;29:1489–520.
Ito M, Okimoto K, Yagura T, Honda G. Induction of sesquiterpenoid production by methyl jasmonate in *Aquilaria sinensis* cell suspension culture. J Essen Oil Res. 2005;17:175–80.
Ishihara M, Tsuneya T, Uneyama K. Components of the volatile concentrate of agarwood. J Essen Oil Res. 1993;5:283–9.
Jayaraman S, Mohamed R. Crude extract of *Trichoderma* elicits agarwood substances in cell suspension culture of the tropical tree, *Aquilaria malaccensis* Lam. Turk J Agric For. 2014;39: 163–73.
Jayaraman S, Daud NH, Halis R, Mohamed R. Effects of plant growth regulators, carbon sources and pH values on callus induction in *Aquilaria malaccensis* leaf explants and characteristics of the resultant calli. J For Res. 2014;25:535–40.

Kazan K, Manners JM. Jasmonate signaling: toward an integrated view. Plant Physiol. 2008;146:1459–68.
Kiyoko M. The book of incense: enjoying the traditional art of Japanese scents. New York: Kodansha International Ltd.; 1992.
Kenmotsu Y, Ogita S, Katoh Y, Yamamura Y, Takao Y, Tatsuo Y, Fujino H, Kadota S, Kurosaki F. Methyl jasmonate-induced enhancement of expression activity of Am-FaPS-1, a putative farnesyl diphosphate synthase gene from *Aquilaria microcarpa*. J Nat Med. 2011;65:194–7.
Konishi T, Konoshima T, Shimada Y, Kiyosawa S. Six new 2-(2-phenylethyl) chromones from agarwood. Chem Pharm Bull. 2002;50(3):419–22.
Kumeta Y, Ito M. Characterization of δ-guaiene synthases from cultured cells of *Aquilaria*, responsible for the formation of the sesquiterpenes in agarwood. Plant Physiol. 2010;154:1998–2007.
Kusari AB, Molina DM, Sabbagh W, Lau CS, Lee B. A conserved protein interaction network involving the yeast MAP kinases Fus3 and Kss1. J Cell Biol. 2004;164(2):267–77.
Li W, Cai CH, Dong WH, Guo ZK, Wang H, Mei WL, Dai HF. 2-(2-phenylethyl) chromone derivatives from Chinese agarwood induced by artificial holing. Fitoterapia. 2014;98:117–23.
Li W, Cai CH, Guo ZK, Wang H, Zuo WJ, Dong WH, Mei WL, Dai HF. Five new eudesmane-type sesquiterpenoids from Chinese agarwood induced by artificial holing. Fitoterapia. 2015;100:44–9.
Lin F, Mei WL, Wu J, Dai HF. GC-MS analysis of volatile constituents from Chinese eaglewood produced by artificial methods. J Chinese Med Mat. 2010;33:222–5.
Liu Y, Chen H, Yang Y, Zhang Z, Wei J, Meng H, Chen W, Feng J, Gan B, Chen X, Gao Z, Huang J, Chen B, Chen H. Whole-tree agarwood-inducing technique: an efficient novel technique for producing high-quality agarwood in cultivated *Aquilaria sinensis* trees. Molecules. 2013;18: 3086–106.
Ma DM, Pu GB, Lei CY, Ma LQ, Wang HH, Guo YW, Chen JL, Du ZG, Wang H, Li GF, Ye HC, Liu BY. Isolation and characterization of AaWRKY1, an *Artemisia annua* transcription factor that regulates the amorpha-4,11-diene synthase gene, a key gene of artemisinin biosynthesis. Plant Cell Physiol. 2009;50(12):2146–61.
Mohamed R, Jong PL, Zali MS. Fungal diversity in wounded stems of *Aquilaria malaccensis*. Fungal Divers. 2010;43:67–74.
Mohamed R, Wong MT, Halis R. Microscopic observation of 'Gaharu' wood from *Aquilaria malaccensis*. Pertanika J Trop Agric Sci. 2013;36(1):43–50.
Mohamed R, Jong PL, Ismail NI. Succession patterns of fungi associated to wound-induced agarwood in wild *Aquilaria malaccensis* revealed from quantitative PCR assay. World J Microbiol Biotechnol. 2014a;30:2427–36.
Mohamed R, Jong PL, Kamziah AK. Fungal inoculation induces agarwood in young *Aquilaria malaccensis* trees in the nursery. J For Res. 2014b;25(1):201–4.
Mucciarelli M, Camusso W, Maffei M, Panicco P, Bicchi C. Volatile terpenoids of endophyte-free and infected peppermint (*Mentha piperita* L.): chemical partitioning of a symbiosis. Microb Ecol. 2007;54(4):685–96.
Naef R. The volatile and semi-volatile constituents of agarwood, the infected heartwood of *Aquilaria* species: a review. Flavour Fragr J. 2011;26:73–89.
Ng LT, Chang YS, Kadir AA. A review on agar (gaharu) producing *Aquilaria* species. J Trop For Products. 1997;2:272–85.
Nimnoi P, Pongsilp N, Lumyong S. Actinobacterial community and diversity in rhizosphere soils of *Aquilaria crassna* Pierre ex Lec assessed by RT-PCR and PCR-DGGE. Biochem Syst Ecol. 2011;39:509–19.
Nobuchi T, Mohd Hamami S. The formation of wood in tropical trees: a challenge from the perspective of functional wood anatomy. Serdang: Penerbit Universiti Putra Malaysia; 2008.
Nobuchi T, Siripatanadilok SA. The formation of wood in tropical forest trees. In: Nobuchi T, Mohd Hamami S, editors. Cytological observations of *Aquilaria crassna* wood associated with the formation of aloes-wood. Serdang: UPM Press; 2008. p. 147–60.
Okugawa H, Ueda R, Matsumoto K, Kawanishi K, Kato A. Effect of jinkoh-eremol and agarospirol from agarwood on the central nervous system in mice. Planta Med. 1996;62:2–6.

Okudera Y, Ito M. Production of agarwood fragrant constituents in *Aquilaria* calli and cell suspension cultures. Plant Biotechnol. 2009;26:307–15.

Oldfield S, Lusty C, MacKinven A. The world list of threatened trees. Combridge, UK: World conservation press; 1998.

Page T, Awarau W. Performance of agarwood (*Aquilaria crassna*) seedling transplants improved by shade and fertiliser. For Ecol Manage. 2012;265:258–69.

Pauli GD, Peart DR, Leighton M, Samsoedin I. An ecological and economic assessment of the nontimber forest product gaharu wood in Gunung Palung National Park, West Kalimantan, Indonesia. Conservation Biol. 2001;15(6):1721–32.

Persoon GA. Agarwood: the life of a wounded tree. IIAS Newsletter. 2007;45:24–5.

Premalatha K, Kalra A. Molecular phylogenetic identification of endophytic fungi isolated from resinous and healthy wood of *Aquilaria malaccensis.* A red listed and highly exploited medicinal tree. Fungal Ecol. 2013;6:205–11.

Pitzschke A, Schikora A, Hirt H. MAPK cascade signalling networks in plant defence. Curr Opin Plant Biol. 2009;12:1–6.

Pojanagaroon S, Kaewrak C. Mechanical methods to stimulate aloes wood formation in *Aquilaria crassna* Pierre ex H Lec (Kritsana) trees. Acta Hort. 2005;676:161–66. doi: 10.17660/ActaHortic.2005.676.20

Qi SY, He ML, Lin DL, Zhang CH, Hu LJ, Zhang HZ. Production of 2-(2-phenylethyl) chromones in cell suspension cultures of *Aquilaria sinensis*. Plant Cell Tiss Org Cult. 2005;83:217–21.

Rahman MA, Khisa SK. Agar production in agar tree by artificial inoculation and wounding. II. Further evidences in favour of agar formation. Bano Biggyan Patrika. 1984;13:57–63.

Rahman MA, Basak AC. Agar production in agar tree by artificial inoculation and wounding. Bano Biggyan Patrika. 1980;9:87–93.

Rodriguez RJ, White JJF, Arnold AE, Redman RS. Fungal endophytes: diversity and functional roles. New Phytol. 2009;182(2):314–30.

Sadgopal, Varma BS. Agar oil from the wood of *Aquilaria agallocha* Roxb. 1952. Indian Forester. 78:26–33.

Santoso E. Agarwood formation by fungal bioinduction technology. In: Susmianto A, Turjaman M, Setio P, editors. Track record: agarwood inoculation technology by FORDA. Bogor: FORDA Press, 2013. 294 p.

Singh B, Sharma RA. Plant terpenes: defense responses, phylogenetic analysis, regulation and clinical applications. 3 Biotech. 2015;5:129–51.

Soehartono T, Mardiastuti A. The current trade in gaharu in West Kalimantan. J Ilmiah Biodiversitas Indonesia. 1997;1:1–10.

Sree N, Udayasri P, Kumar Y, Babu B, Kumar Y, Varma M. Advancements in the production of secondary metabolites. J Nat Prod. 2010;3:112–23.

Takemoto H, Ito M, Shiraki T, Yagura T, Honda G. Sedative effects of vapor inhalation of agarwood oil and spikenard extract and identification of their active components. J Nat Med. 2008;62:41–6.

Tamuli P, Boruah P, Nath SC, Leclercq P. Essential oil of eaglewood tree: a product of pathogenesis. J Essen Oil Res. 2005;17:601–4.

Thanh LV, Do TV, Son NH, Sato T, Kozan O. Impacts of biological, chemical and mechanical treatments on sesquiterpene content in stems of planted *Aquilaria crassna* trees. Agroforest Syst. 2015;89:973–81.

Tawfik HA, Ewies EF, El-Hamouly WS. Synthesis of chromones and their applications during the last ten years. Int J Res Pharm Chem. 2014;4(4):1046–85.

TRAFFIC East Asia-Taipei and TRAFFIC Southeast Asia. The trade and use of agarwood in Taiwan, Province of China. CITES; 2005.

TRAFFIC Oceania. Convention on International Trade in Endangered Species of Wild Fauna and Flora. Agarwood (Gaharu) harvest and trade in new Guinea [Papua New Guinea and the Indonesia province of Papua]. Twelfth meeting of the Plants Committee, CITES Decisions 11.112 and 11.113 regarding *Aquilaria* spp. Leiden; 2002. May 13–17.

Turjaman M, Santoso E. Progress research: agarwood cultivation and inoculation technology in Asia. In: Tampubolon AA, et al., editors. National seminar : non-timber forest products. Mataram, Lombok: Forest Research Institute for NTFPs; 2012. 500 p.

Wang JW, Wu JY. Effective elicitors and process strategies for enhancement of secondary metabolite production in hairy root cultures. Adv Biochem Eng Biotechnol. 2013;134:55–89.

Wang AM, Zhang FK, Huang LF, Yin XP, Li HF, Wang QY, Zeng ZW, Xie T. New progress in bio-catalysis and biotransformation of flavonoids. J Med Plants Res. 2010;4(10):847–56.

Wang SL, Hwang TL, Chung MI, Sung PJ, Shu CW, Cheng MJ, Chen JJ. New flavones, a 2-(2-phenylethyl)-4H-chromen-4-one derivative, and anti-inflammatory constituents from the stem barks of *Aquilaria sinensis*. Molecules. 2015;20(11):20912–25.

Wei JH, Yang Y, Zhang Z, Meng H, Feng JD, Gan BC. Liquid infusion method for producing linaloe on *Aquilaria sinensis* trees China Patent. CN101755629A; 2010.

Wijitphan P. Method to stimulate resin formation by wounding on the *Aquilaria*'s trunk. US Patent. US7485309–B1; 2009.

Wu B, Lee JG, Lim CJ, Jia SD, Kwon SW, Hwang GS, Park JH. Sesquiterpenoids and 2-(2-phenylethyl)-4H-chromen-4-one (=2-(2-phenylethyl)-4H-1-benzopyran-4-one) derivatives from *Aquilaria malaccensis* agarwood. Helv Chim Acta. 2012;95:636–42.

Wyn LT, Anak NA. Wood for the trees: a review of the agarwood (gaharu) trade in Malaysia. Petaling Jaya/Selangor/Malaysia: TRAFFIC Southeast Asia; 2010.

Xu YH, Wang JW, Wang S, Wang JY, Chen XY. Characterization of GaWRKY1, a cotton transcription factor that regulates the sesquiterpene synthase gene (+)-δ- cadinene synthase-A. Plant Physiol. 2004;135:507–15.

Xu Y, Zhang Z, Wang M, Wei J, Chen H, Gao Z, Sui C, Luo H, Zhang X, Yang Y, Meng H, Li W. Identification of genes related to agarwood formation: transcriptome analysis of healthy and wounded tissues of *Aquilaria sinensis*. BMC Genomics. 2013;14:1–16.

Yagura T, Shibayama N, Ito M, Kiuchi F, Honda G. Three novel diepoxy tetrahydrochromones from agarwood artificially produced by intentional wounding. Tetrahedron Lett. 2005;46:4395–8.

Yang DL, Wang H, Guo ZK, Li W, Mei WL, Dai HF. Fragrant agarofuran and eremophilane sesquiterpenes in agarwood 'Qi-Nan' from *Aquilaria sinensis*. Phytochem Lett. 2014;8:121–5.

Yoshii E, Koizumi T, Oribe T, Takeuchi F, Kubo K. The structure of agarotetrol, a novel highly oxygenated chromone from agarwood (jinko). Tetrahed Lett. 1978;19:3921–4.

Yu ZX, Li JX, Yang CQ, Hu WL, Wang LJ, Chen XY. The jasmonate responsive AP2/ERF transcription factors AaERF1 and AaERF2 positively regulate artemisinin biosynthesis in *Artemisia annua* L. Mol Plant. 2012;5:353–65.

Zhang XL, Liu YY, Wei JH, Yang Y, Zhang Z, Huang JQ, Chen HQ, Liu YJ. Production of high-quality agarwood in *Aquilaria sinensis* trees via whole-tree agarwood-induction technology. Chin Chem Lett. 2012;23:727–30.

Zhang Z, Han XM, Wei JH, Xue J, Yang Y, Liang L, Li XJ, Guo QM, Xua YH, Gao ZH. Compositions and antifungal activities of essential oils from agarwood of *Aquilaria sinensis* (Lour.) Gilg induced by *Lasiodiplodia theobromae* (Pat.) Griffon. & Maubl. J Braz Chem Soc. 2014a;25:20–6.

Zhang Z, Wei J, Han X, Liang L, Yang Y, Meng H, Xu Y, Gao Z. The sesquiterpene biosynthesis and vessel-occlusion formation in stems of *Aquilaria sinensis* (Lour.) Gilg trees induced by wounding treatments without variation of microbial communities. Int J Mol Sci. 2014b;15: 23589–603.

Chapter 4
Development of Agarwood Induction Technology Using Endophytic Fungi

Maman Turjaman, Asep Hidayat, and Erdy Santoso

Abstract Agarwood plays an important role in gaining foreign exchange and as a source of income for people living in, around, and inside the forest of Indonesia. Its production has declined rapidly due to the lack of proven technology for its induction. If no serious action is taken now, agarwood production will not be sustainable in the future. Agarwood formation is initiated by biotic and/or abiotic factors and is affected by host tree, microbe, and environment. Nonetheless, in cultivating agarwood on a large scale, a standard operating procedure is the principal factor that determines agarwood quantity and quality. Realizing the importance of a procedure for agarwood induction, we developed a technology using the biotic factor and its natural way in forming agarwood. We applied selected endophytic fungi directly on *Aquilaria* and *Gyrinops* trees. Because fungi are living organisms, naturally they help to spread the induction mechanism into other regions of the tree. Remarkably, the technology produced substantial amount of agarwood of high-grade quality. Although a better selection of fungal strains must be discovered and tested in field trials, this promising technology could be the answer to commercial agarwood production in Indonesia and could be adopted by other countries.

4.1 Introduction

For most of the forest communities in Indonesia, agarwood products have been known to retain social, cultural, and economic values (Donovan and Puri 2004). In other communities, it has been used as raw ingredients in fragrance, aromatherapy, pharmaceutical, and herbal medicines for centuries, such as in Buddhist, Hindu, and

M. Turjaman (✉) • E. Santoso
Forest Microbiology Research Group, Forest Research and Development Centre (FRDC), Environment and Forestry Research, Development, and Innovation Agency (FORDA), Ministry of Environment and Forestry, Jalan Gunung Batu No. 5, Bogor 16680, Indonesia
e-mail: turjaman@gmail.com

A. Hidayat
Forest Plantation Fiber Technology Research Institute, FORDA, Ministry of Environment and Forestry, Kuok, Riau Province, Indonesia

R. Mohamed (ed.), *Agarwood*, Tropical Forestry,
DOI 10.1007/978-981-10-0833-7_4

Islamic cultures (Heuveling van Beek et al. 1999; Barden et al. 2000). In China, the Chinese agarwood, *Aquilaria sinensis*, is used as a crude drug in traditional sedative, analgesic, and digestive medicine (Yang et al. 2013). The agarwood-producing tree species are widely distributed in Indonesia, including Sumatra, Borneo, Java, Nusa Tenggara, Sulawesi, Maluku, and Papua islands. High economic values and high demand have triggered overexploitation of the species in their natural habitats. The increasing rate of overexploitation is unfortunately not being replaced by proper cultivation technique; this highly threatens the species. Consequently, two important agarwood-producing genera, *Aquilaria* and *Gyrinops*, have been classified as threatened, which need proper regulation and restriction in their harvesting (CITES 2005).

A new trend in the agarwood industry termed as "one-stop service" has been on the rise in agarwood-producing countries. Based on this new scheme, the production chain of agarwood is managed comprehensively from upstream to downstream. Through this scheme, agarwood will not be sold exclusively as a raw material, but also processed into various kinds of agarwood-based products according to market demands like perfume, incense, soap, tea, and herbal medicine. Currently, the world's demand for agarwood products reaches 4500 tonnes per year and it is predicted to keep increasing year after year (Santoso 2015). To meet the demand, agarwood-producing countries in Indochina including Laos, Cambodia, Vietnam, and Thailand have started establishing their agarwood industry. Countries with long history of agarwood-bounded communities for many generations seem to be more equipped in managing comprehensive agarwood cultivation and production; they started from massive planting of the species to establishing research-based agarwood industry. Attention is given to every stage in the setting up of a comprehensive agarwood plantation including developing proper cultivation and inoculation techniques to maintain product quantity and quality. Generally, big agriculture companies, whose product segmentation had been correctly determined from the beginning, own such plantations. The companies will execute their management plans to fulfill their targets (Santoso 2015).

Many studies have been conducted on agarwood by researchers from many countries including those in Asia, Europe, and America. However, the results are scarce in the public domain and tend to stay as internal secrets, available only to limited communities. The high commercial value of agarwood and its derivatives had triggered such research to be patented and become inaccessible to the public. Many workshops, meetings, and seminars, both at international and national levels, mostly present the agarwood research in general terms and introduce only bits of information of their patented findings.

Agarwood is a scented product obtained after a pathological situation happened to the wood of the standing trees of the family Thymelaeaceae (most commonly the *Aquilaria* genus). More specifically, the condition could be initiated by an endophytic fungal invasion. Other fungi such as molds and decay fungi might play a role as well. Among the different fungal species associated with agarwood formation, few could exhibit pathogenesis, while others seem to be saprophytic (Tamuli et al. 2008). Indeed *Aquilaria* is naturally infected by a variety of fungi including

Aspergillus sp., *Botryodiplodia* sp., *Diplodia* sp., *Fusarium bulbiferum*, *F. laterium*, *F. oxysporum*, *F. solani*, *Penicillium* sp., and *Pythium* sp. (Sitepu et al. 2011). The involvement of these biotic agents perhaps explains the slow formation process of this chemically laden product in the natural habitat. In order to meet the demand for agarwood and protect the wild *Aquilaria* trees, many countries with *Aquilaria* plantations also include research on agarwood cultivation and inoculation techniques in their strategy.

In general, there are three approaches used to develop agarwood inoculation techniques: physical-mechanical, chemical, and biological. Chemical-based inoculation technology has been developed and practiced in Indonesia, Malaysia, Thailand, Cambodia, Vietnam, Laos, and China (Santoso 2015). However, some have no guarantee of yielding commercial agarwood product such as in a case reported in Indonesia, where farmers had used some chemical substances to stimulate the process of agarwood formation, only to find that the technique is ineffective. Unfortunately, they lost 6–7 years of their time in maintaining the trees before they learned that their efforts failed to produce agarwood. Agarwood cultivation should be developed based on its natural formation. The philosophy of agarwood formation stands as a triangle connecting the tree, microbes, and the environment. By understanding interaction of each unit in the triangle, it will be possible to apply endophytic fungi as inoculants into the host tree to yield commercial agarwood in the future. This chapter reviews and analyzes the utilization of artificial inoculation in the induction process of agarwood practiced by several countries in Asia with special reference to Indonesia.

4.2 Agarwood-Producing Species in Indonesia

Indonesia has a diverse collection of naturally growing agarwood-producing species when compared to other countries in Asia. Based on extensive ground surveys, farmers in the islands of Sumatra, Kalimantan, Java, Sulawesi, Nusa Tenggara, Moluccas, and Papua have been cultivating agarwood trees of at least seven different species. However, the plantation size is small, ranging from 10 to 5000 trees per farmer. A different trend is observed in neighboring countries such as Malaysia, Thailand, Vietnam, Laos, Cambodia, and China, where the plantation areas are of larger scale, starting as low as 40 hectares to more than 1000 hectares.

Among the species planted in Indonesia, *Aquilaria malaccensis* also known as "malakensis" is the most favorite species for mass cultivation in western Indonesia. Other species in the *Aquilaria* genus such as *A. microcarpa*, *A. beccariana*, and *A. hirta* have also gained popularity among farmers in Sumatra and Kalimantan. Another competitor of *Aquilaria*, also producing agarwood, is *Gyrinops*. Members of this genus are planted in the eastern part of Indonesia and are known by the trade name "filaria." Generally, the name "filaria" is being loosely used on other species as well, such as on *Aquilaria filaria*, *Aquilaria cumingiana*, *Gyrinops versteegii*, and other *Gyrinops* species. The most important "filaria" species is *G. versteegii*. The species is planted in eastern Indonesia, covering several islands including the Lesser Sunda Islands (Lombok, Sumbawa, Flores, Sumba), North Celebes (Minahasa), and Papua. Although it is the most popular species in eastern Indonesia, *G. versteegii* is less favored compared to *A. malaccensis* when it comes to agarwood cultivation. Agarwood grouped under the trade names "filaria" and "malakensis" has no actual significance on the agarwood's quality. This grouping was applied merely to indicate the origins of the trees: "filaria" (eastern) and "malakensis" (western). In fact, both groups are included in CITES Appendix II for its regulation system in the international trade. To regulate agarwood trade, a DNA fingerprint database based on microsatellites has been set up for *Aquilaria crassna* to detect the geographic origins of traded wood and incense samples for forensic applications (Eurlings et al. 2010). This method could be adopted for detecting the geographic origins of *Aquilaria* and *Gyrinops* species in Indonesia.

4.3 Harvesting Natural Agarwood in Indonesia

Indonesian natural agarwood has a long history and is known worldwide. It was recorded as the main commodity bartered between the empires of China and the kingdoms of Indonesia since the Silk Road era. Traditionally, natural agarwood is gathered by cutting down the infected trees. Agarwood hunters usually have their own parameters in determining a suspected tree, among them are fallen yellow leaves and presence of black ants entering and exiting the tree trunk through small holes. For example, the indigenous people of East Kalimantan rely on several phenomena as external indicators of agarwood presence inside a tree. Symptoms of infection recognized by the local people include insect bore holes, knots, a hollow sound upon thumping, tumorlike growths, bark drop, and excessive leaf fall (Donovan and Puri 2004). To harvest the agarwood, hunters usually chipped all infected trunks, stems, and branches. Indonesia exported high-quality agarwood in small quantities before the 1990s. However, as the market demand keeps increasing, hunters tend to harvest natural agarwood in a speculative manner causing the depletion of agarwood-producing tree stocks in the nature. Hunters usually have their upper ordinate whose role is to support them financially. The fundraiser usually gives some amount of money as a disbursement to the hunters for their agarwood hunting trips in the forest, which can last between 1 and 3 months. By the time the hunters found the agarwood, the product price will be fully determined by their fundraiser. In most cases, there will be no bargaining position for the hunters. In this monopsony system, the hunters will not likely gain as much profits as the fundraiser (Siran 2013).

At the end of 1970s, demand for agarwood had increased significantly because the supply of high-quality resin wood from Cambodia and Vietnam diminished due to the political situation. At the same time, Saudi Arabia and the Gulf Emirates experienced the oil boom, which generated high income and consequently an increase in the demand for agarwood. Arab people from the neighboring Middle

Eastern countries also benefited from the oil boom and spent more on luxury products such as agarwood (Gunn et al. 2004). Between the 1980s and early 1990s, the "agarwood fever" had surfaced in East Kalimantan, where expeditions of professional collectors, sometimes dropped by helicopters and sponsored by bogus traders, were organized to hunt for agarwood. At the time, about 70 % of the collected agarwood was exported to the Middle East, while the rest to China, Hong Kong, Taiwan, and Japan. By 1995, traders stopped funding high-cost expeditions in Kalimantan and turned instead to the Papua island. There is a positive correlation between the numbers of felled trees from the wild and the rate of agarwood trade. Considering the high demand for agarwood products, the current population density of *A. malaccensis* in Indonesia has lowered to only 1–2 individuals per hectare. Although seed production was reportedly high, seed dispersal is limited and germination rate is low. Other factors that caused depletion of the agarwood tree populations in the wild include forest fires, illegal logging, forest conversion, and mining concession (Soehartono and Newton 2000, 2001a).

The formation of natural agarwood is a complex process (Santoso 2013). In the wild, it is usually triggered by wounding of twigs or branches due to friction between trees, or caused by wind, thunder, people, or wild animals. Borer insects and caterpillars could also make a hole in the tree and trigger the formation. Wounding tissues could easily be infected since fungi are present abundantly in nature in the form of spores or hyphae and are dispersed through water, wind, and soil. Suitable environment conditions such as high humidity and availability of carbon and energy sources can boost fungal growth. Fungi often secrete poisonous or toxic compounds during the infection process and the tree reciprocates by releasing chemical compounds as one of the defense mechanisms to counter the attack. The mechanism occurs continuously and this is believed to be the key process in agarwood formation. As happened in the wild, the process usually takes many years to produce high-quality natural agarwood, maybe even hundreds of years. The slow process explains why products from natural agarwood are highly valuable and fetch high prices in the market.

4.4 Establishment of Agarwood Plantation in Asia

Agarwood-producing trees are widely distributed in Asia, including India, Sri Lanka, Bangladesh, Bhutan, Myanmar, Laos, Vietnam, Cambodia, China, Brunei, Malaysia, Indonesia, Philippines, Thailand, and Papua New Guinea. There are more than 20 species within the genus *Aquilaria* and less than 10 species within the genus *Gyrinops*. All these species have been listed in CITES Appendix II since 2005 (COP 13). Among them, *A. malaccensis* and *G. versteegii* are most popular in Indonesia. These species, as well as *A. crassna* and *A. filaria*, are regarded as agarwood producers of commercial quality. Due to the economic value of these species, translocation or transplanting across different islands and countries became common. Such as the case with *A. crassna* of Indochina origin; this species has been planted widely across the regions in Sumatra.

In Asia, the establishment of an agarwood plantation usually originates from generative propagation by collecting seeds and germinating them in the nursery. Normally, the fruiting season happens twice a year. Trees that fruit during the months of July to August will mature in November to December, while those that fruit between March and April will mature in July to August. Most farmers in Indonesia are still managing their small-scale plantation traditionally. They are dependent on the wildings to supply their planting stocks. New micropropagation or tissue culture technique to fulfill the need for planting stocks is still untried. Thailand and Laos are among the few countries that use micropropagation as their propagation technique for setting up commercial plantations. Seedlings of *A. crassna* propagated by tissue culture in Thailand have been exported to northern Australia. However, such technique is more expensive when compared to collecting wildings. Moreover, tissue culture requires an investment for laboratory facilities and skilled human resources. Regardless of its high cost, mass vegetative propagation by means of tissue culture should be considered when superior mother trees are available as explant sources.

Field visits to several locations in different Asian countries revealed that agarwood-producing plantations have been practicing standard silviculture technique in many aspects, starting from nursery management to field planting and post-harvesting. In Vietnam, the agarwood association maintained that they have planted more than 20 million agarwood-producing trees in established plantations. In Cambodia, an agarwood farmer can plant as many as 2 million trees. Most companies or private farmers usually plant species that are already well known for their superior genetic characteristics such as with *A. malaccensis* and *A. crassna*. For the same reason, these two species are widely planted in many countries. In the near future, breeding strategies of such species will be in high demand to produce quality-planting materials that can yield agarwood of "super" or "double super" quality.

4.5 Agarwood Induction Technology

Generally, it is agreed that this aromatic resin is produced as the tree sap thickens in response to wounding and fungal infection (Donovan and Puri 2004). The resinous agarwood acts as a chemical barrier to attacks by fungi and insects, which otherwise is not formed by the tree (Paoli et al. 2001). Agarwood formation takes place in different woody organs like in the stem, branch, and roots, with one condition, they must be infected. Due to the importance of infection in inducing agarwood, induction techniques have been developed and are discussed below.

4.5.1 Physical-Mechanical Inoculation

Physical-mechanical inoculation has been practiced since long time ago. Agarwood farmers in different Asian countries have tried several wounding methods to produce agarwood, including ax chopping, nailing, and holing. These methods often take a long time, with generally poor yield and low quality in the agarwood they produced (Liu et al. 2013). In Indonesia, these methods are widely practiced in Riau-Sumatra, East Kalimantan, and Lombok. The initial process is tree wounding, usually performed by a blade, knife, cleaver, and spikes. Naturally, the wounded tissue will be infected by fungi, and this leads to the gradual formation of agarwood. This technique is highly risky and very speculative because the amount of agarwood produced is unpredictable, often in small amounts.

Another type of mechanical inoculation is by imbedding nails into the tree, lengthwise. In average, a mature wild tree supposedly will need around 20 kg of nails if the nail is to be imbedded every 10 cm along the length of the tree. Farmers believed that trees treated by this practice can be harvested at most 2 years after the treatment. However, the produced agarwood has been found of low quality and is frequently tainted by blue stains. Furthermore, the scent is inferior due to contamination from the smell of rusty nails. The use of nails is only applicable for a small number of trees as it will be less efficient and cost ineffective if applied in a big company plantation, especially when labor wages in Southeast Asian countries are starting to rise. At several locations in Sumatra and Kalimantan, heated nails have also been used. Similar mechanical inoculation technique is also being practiced in Bangladesh on *A. malaccensis* and in Yunnan Province (P.R. China) on *A. sinensis* and *Aquilaria yunnanensis*. Similar to those in Indonesia, this inoculation technique produces small amounts of agarwood with low-grade quality. Other examples of mechanical inoculation are by scratching the tree trunk with sharp tools such as a machete or a cleaver, or by debarking and sawing the stem. Nonetheless, the quantity and quality of agarwood produced through these techniques are still without certainty.

4.5.2 Chemical Induction

Chemical induction technique has been applied in many sourcing countries. Jasmonic acid, sulfuric acid, acetic acid, and alcohol are among the common chemical compounds applied to initiate agarwood formation. Jasmonic acid has been proven in Vietnam; it induced agarwood formation to the thickness of 2–3 mm. The activation of agarwood tree secondary to metabolism by jasmonates can be divided into two modes: (1) the augmentation of biosynthetic activities constitutively observed in plants and (2) the induction of latent biosynthetic abilities that occur only during unusual physiological processes such as mechanical wounding and microbial infection (Kenmotsu et al. 2013). Sulfuric and acetic acids have also been practiced in several countries; however, the results were not so successful, and in some cases, the trunks were broken.

Some chemicals can be extremely toxic to human. Therefore, the choice of chemical is of paramount importance especially when the agarwood is intended for use as raw materials in making perfumes, incenses, tea, and medicines. However, agarwood formed by this technique is not completely useless because after elimination of the toxic compounds, the product can still be used for other purposes such as decoration materials and accessories. When compared to mechanical induction, chemical induction appears to be a faster process, whereby the wood color changes from white to blackish in matters of days to weeks. Delivery methods of the chemicals have also been tested. The most popular method uses a needleless syringe that is attached to a tubing that connects to a reservoir containing the chemical solution. The chemical solution is then taken up from the point of syringe insertion, usually at the bottom of the stem, to the upper parts of the tree. After some time, the wood color will change from whitish to dark brown. Other delivery techniques are like the bamboo sticks, which are soaked into the chemical solution and then spiked into the trunks and branches. This practice was conducted by several groups of farmers in Indonesia; however, it is now becoming unpopular due to its side effect. The chemicals injected into the trees could be released back to the environment and cause water and soil pollution.

4.5.3 Biological Inoculation

Many researchers believe that microbes have an important role in agarwood formation (Tamuli et al. 2005; Mohamed et al. 2010). Since early 1990s, biological inoculation using specific fungi as inoculants has been practiced. However, wounding has to precede inoculation because it is the entry mode for the fungi to infect the host and initiate agarwood formation. Wounding can occur naturally from the act of animal or other physical interactions between the tree and its environment. Introduction of fungal propagules such as spores into the wounded tissue causes the host tree to secrete a combination of metabolic compounds. Major metabolic compounds reported from infected woods of *A. malaccensis* are chromones, aromatic compounds, sesquiterpenes, monoterpenes, sterols, and fatty acid methyl esters. Endophytic fungi have been associated to important agarwood compounds when used as inoculants and can be detected as early as 6 months after inoculation (Jong et al. 2014). However, not all fungi can trigger agarwood formation on the host tree. Some endophytic fungi are able to change the color of the wood fiber and cause the presence of the fragrant resin on the host tree. *Acremonium* sp., for example, could change the color of wood from white to dark brown, but the scent of agarwood appears to be inconsistent. More efforts have to be carried out to determine the most suitable agarwood-inducing endophytic fungi. As reported by the local Penan people in East Kalimantan, an insect may be implicated in the etiology of agarwood formation. Insect often bores through the cracks and crevices in the bark, while carrying fungal propagules to the underlying wood (Donovan and Puri 2004). Using insects as the vector for transmitting endophytic fungi (e.g., weevils, termites) could also be applied to accelerate agarwood formation. However, control on insect population should be determined strictly to prevent outbreak that can negatively affect the trees.

4.6 Endophytic Fungi-Based Inoculation

Agarwood is formed in the main stem, branches, and roots where natural wounding usually occurs. Agarwood-producing trees can be exemplified as a warehouse for various microbes, commonly endophytic fungi. Endophytes are microorganisms that maintain endosymbiotic relationship within plants at least in one stage of their life cycle. Both fungi and plants will receive ecological benefit from this relationship. Some genera of known endophytic fungi include *Alternaria*, *Cladosporium*, *Curvularia*, *Fusarium*, *Phaeoacremonium*, and *Trichoderma*. In the process of infection, fungi secrete several pathogenesis-related enzymes such as polyphenol oxidase, peroxidase, pectinase, and cellulase. These enzymes have been detected in naturally infected and inoculated *A. malaccensis*. Samples infected naturally with *Chaetomium globosum* and *F. oxysporum* exhibit higher activity of all the enzymes when compared to healthy samples (Tamuli et al. 2008). Changes in the activity of various enzymes in naturally infected tree indicate that they may be involved in the infection process and development of disease symptoms in agarwood trees. Many scientists believe that wounding causes the tree to weaken and then become vulnerable to attacks by a pathogenic fungus. Other factors such as tree age, genetic background, and seasonal and environmental variation may be important in agarwood formation (Ng et al. 1997). The complex process of natural agarwood formation has attracted many scientists to conduct extensive research on this phenomenal natural product. Scientists believe that the triangle factors, host tree, endophytic fungi, and environment, are important players of the formation process (Fig. 4.1).

4.6.1 *Host Tree*

The host tree is confined to the family Thymelaeaceae. There are six genera in this family: *Aquilaria*, *Gyrinops*, *Enkleia*, *Gonystylus*, *Wikstroemia*, and *Aetoxylon*. Some of the species found in Indonesia are listed here; however, not all have been examined for their agarwood-producing abilities: *A. hirta*, *A. filaria*, *A. malaccensis*,

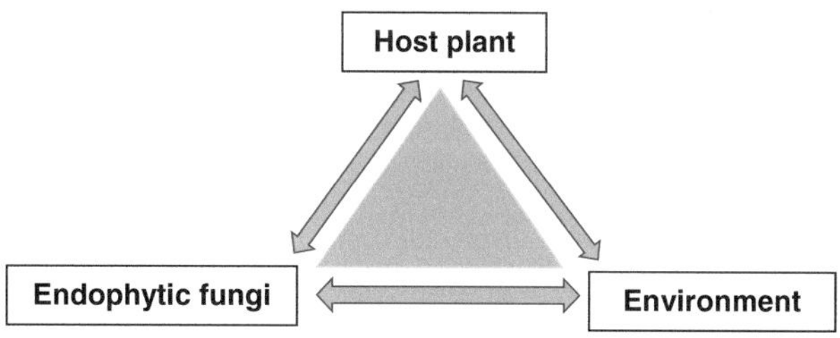

Fig. 4.1 Agarwood is formed under the influence of the plant disease triangle paradigm (genetic of the host tree, pathogenic endophytic fungi, and conducive environment), which affects the amount and quality of agarwood produced in a host tree (Santoso 2013)

A. microcarpa, *A. beccariana*, *A. cumingiana*, *G. versteegii*, *G. moluccana*, *G. decipiens*, *G. ledermannii*, *G. salicifolia*, *G. podocarpa*, *Enkleia malaccensis*, *Gonystylus bancanus*, *Wikstroemia polyantha*, *W. tenuriamis*, *W. androsaemofilia*, and *Aetoxylon sympetalum* (Sitepu et al. 2011). Because the taxonomy and systematics of this family have not been resolved, several of the names listed here could be synonyms or misidentified cases. A comprehensive study is still needed to classify the species found in Indonesia. Nevertheless, among the Thymelaeaceae members, *Aquilaria* and *Gyrinops* are the most well-known high-grade agarwood-producing genera in the world.

4.6.2 Environmental Factors

There are several environmental factors that influence agarwood formation. They include soil fertility, temperature, humidity, light intensity, and pests and diseases (Pratiwi et al. 2011; Purnomo and Turjaman 2011). Indeed, many defense strategies have evolved to fight off these numerous factors (Wong et al. 2013). For example, in a field study at Bangka Belitung province, Indonesia, trees grown on poor soil happened to yield high-grade agarwood when compared to trees grown on rich soil (Pratiwi et al. 2011). Agarwood hunters in Gunung Palung (West Kalimantan) observed that the occurrence of agarwood is more frequent in trees at high elevations and on poor soils, and trees growing under stressful conditions are more vulnerable to infection (Paoli et al. 2001). The occurrence of pests and diseases on a tree also influences inoculation procedure. Trees that are heavily under attack by pests (e.g., *Heortia vitessoides*) are not recommended for endophytic fungal inoculation. This is because the trees are already sick and inoculation with an endophytic fungi will only worsen the trees' condition possibly leading to death. Previously attacked trees can be inoculated after the trees are allowed some times to recover from the attack. Planting essential oil-producing species such as *Citronella* could help in reducing the occurrence of pests and diseases. *Citronella* has some aromatic attractants that can protect agarwood-producing trees from the attacking females of *H. vitessoides*.

4.6.3 Endophytic Fungi

The wounded trunk of an agarwood-producing tree in the natural environment harbors multiple fungal taxa that exist in a complex system as a whole or in succession leading to agarwood production in the tree trunk (Mohamed et al. 2010, 2014). Isolation of many endophytic fungi from agarwood trees have been carried out extensively around the world since 1934 (Sitepu et al. 2011). It is unlikely that a single fungal species could be responsible for agarwood. More than 20 fungal species have been identified in the aromatic wood and the number keeps increasing. Fungi of several genera have been determined as agents in agarwood formation including *Torula* sp., *Aspergillus* sp., *Fusarium* sp., *Lasiodiplodia* sp., *Penicillium* sp., and many more. Research carried out by a team at the Environment and Forestry, Research, Development, and Innovation Agency (FORDA), Ministry of Environment and Forestry, Indonesia, identified *Fusarium solani* as one of the most effective agents in agarwood formation.

The capability to induce agarwood is different for each fungal species. Some are slow to infect and induce agarwood, while others are more aggressive. Thus, isolation and screening for appropriate fungi are necessary to assure successful induction. Being aware of the importance of endophytic fungi to the Indonesian agarwood industry, FORDA has taken the initiative to study and characterize fungal isolates collected from all over Indonesia (Table 4.1) (Sitepu et al. 2011). The cultures are deposited at the Indonesian Tropical Forest Culture Collection (INTROF-CC) to ensure their viability and for future use.

Agarwood cultivation requires a quite fair amount of capital due to its complex process. Firstly, wild agarwood seedlings must be planted and maintained for 6–7 years. Then, trees that are big enough (>20 cm in diameter) require artificial induction to produce agarwood. Following inoculation is monitoring the trees until the time to harvest. The optimal time for harvesting is 3 years after inoculation upon which the agarwood usually has reached maturity. Finally, the cleaning and carving process take place to separate the agarwood from non-agarwood parts. This process is done manually hence it is labor intensive (Mucharommah 2011).

Our research series determined that the endophytic fungi application that we developed at FORDA is a promising technology that can yield high-grade agarwood product (Santoso 2015). The inoculation procedure is easy to conduct by forest communities. The procedure begins by selecting a tree with diameter at breast height (dbh) of more than 15–20 cm. Then holes of 3 mm in diameter are made in a spiraling pattern along the stem starting with the first hole at the trunk base (point 1) (Fig. 4.2a). Then the second hole is made 7–10 cm in the horizontal direction and 12–20 cm in the vertical direction (point 2). This pattern is repeated until the highest reachable shoot tip. Each tree of 15–20 cm dbh has about 100–200 holes. The maximum depth a wound can be drilled is one-third of the stem diameter. Once ready, each hole is inoculated with one cc of the inoculum, which is in liquid form (Fig. 4.2b). The inoculation procedure requires a skilled and courageous personnel, who is physically fit to work in the field. After 2–3 months, agarwood formation can be detected and evaluated. Using this technology, we have inoculated many trees in different provinces in Indonesia and the results are very promising (Fig. 4.3a–c). An *A. malaccensis* tree in Sanggau (West Kalimantan) with 15 cm dbh yielded 4.5 kg of dried weight (dw) agarwood, with the selling price of about USD 200 per kg. In Kandangan (South Kalimantan), another *A. malaccensis* (40 cm dbh) yielded 13 kg (dw) of agarwood, 18 months after inoculation with *F. solani* (Turjaman and Santoso 2012). With advancement in the inoculation technology, it is not surprising that within the next 5 years, cultivated agarwood of various qualities will be released into the market. This trend will help to ease the tension on natural agarwood exploitation.

Table 4.1 A list of 36 strains of fungal endophytes associated to agarwood identified through molecular methods

No.	Isolate number	Origin (province)	Molecular identification
1.	FORDA-CC 506	North Sumatra	*Fusarium solani*
2.	FORDA-CC 509	Gorontalo	*Fusarium solani*
3.	FORDA-CC 503	West Sumatra	*Fusarium solani*
4.	FORDA-CC 512	Papua	*Fusarium solani*
5.	FORDA-CC 500	Jambi	*Fusarium solani*
6.	FORDA-CC 501	West Sumatra	*Fusarium solani*
7.	FORDA-CC 510	Molluca	*Fusarium solani*
8.	FORDA-CC 497	Central Kalimantan	*Fusarium solani*
9.	FORDA-CC 499	West Kalimantan	*Fusarium solani*
10.	FORDA-CC 2372	East Nusa Tenggara	*Fusarium solani*
11.	FORDA-CC 504	Riau	*Fusarium solani*
12.	FORDA-CC 514	Papua	*Fusarium solani*
13.	FORDA-CC 502	West Sumatra	*Fusarium ambrosium*
14.	FORDA-CC 515	East Nusa Tenggara	*Fusarium* sp.
15.	FORDA-CC 2379	Molluca	*Fusarium solani*
16.	FORDA-CC 511	West Nusa Tenggara	*Fusarium solani*
17.	FORDA-CC 2370	Bangka Belitung	*Fusarium solani*
18.	FORDA-CC 517	Bangka Belitung	*Fusarium solani*
19.	FORDA-CC 513	Papua	*Fusarium solani*
20.	FORDA-CC 519	West Java	*Fusarium falciforme*
21.	FORDA-CC 2375	East Kalimantan	*Fusarium oxysporum*
22.	FORDA-CC 520	West Java	*Fusarium solani* f. batatas
23.	FORDA-CC 518	Bangka Belitung	*Fusarium solani* f. batatas
24.	FORDA-CC 2371	Bangka Belitung	*Fusarium solani*
25.	FORDA-CC 2377	West Java	*Fusarium solani*
26.	FORDA-CC 507	Lampung	*Fusarium solani* f. batatas
27.	FORDA-CC 498	Central Kalimantan	*Fusarium solani*
28.	FORDA-CC 2369	West Sumatra	*Fusarium ambrosium*
29.	FORDA-CC 495	South Kalimantan	*Fusarium solani*
30.	FORDA-CC 2373	West Nusa Tenggara	*Fusarium solani* f. batatas
31.	FORDA-CC 2374	East Kalimantan	*Fusarium solani*
32.	FORDA-CC 508	Bengkulu	*Fusarium* sp.
33.	FORDA-CC 505	North Sumatra	*Fusarium solani*
34.	FORDA-CC 496	South Kalimantan	*Fusarium solani* f. batatas
35.	FORDA-CC 516	Bangka Belitung	*Fusarium solani*
36.	FORDA-CC 2378	West Java	*Fusarium solani*

The isolates were collected from 17 provinces in Indonesia (Sitepu et al. 2011)

Fig. 4.2 Application of the fungal inoculation technology on an agarwood-producing tree at a community forest. (**a**) A hole is made using a 3 mm drill bit, and (**b**) a total of 1 cc liquid inoculum is applied into the hole

Fig. 4.3 Agarwood was successfully produced using the endophytic fungal inoculation technology developed at FORDA. Some examples of the produced agarwood: (**a**) after 24 months from South Sumatra, and (**b**) and (**c**) after 15 months from West Kalimantan

4.7 Concluding Remarks

Agarwood bioinduction technology by means of endophytic fungi is proven to be a promising technique for agarwood formation. This technique forms agarwood at a faster rate compared to customary techniques, is more environmentally friendly compared to chemical inducers, and is widely acceptable from health and safety perspectives compared to physical or chemical methods. This method was developed following natural agarwood formation mechanism, whereby wounds were first initiated by making holes on the trunk followed by inoculating a specific fungal endophyte into the "wound." Triangular factors, namely, the host tree, microbes, and environment, are the important elements in determining a successful formation of agarwood. Selecting appropriate tree species, inoculating a suitable fungus, and establishing supporting environmental factors will yield agarwood of commercial value. Applying an accurate and proven inoculation technique in the field should become a serious concern to support the success of any agarwood cultivation program. Our comprehensive inoculation technique, developed since 1984, has proved that establishing a manmade agarwood plantation is a prospect that could materialize very soon. Nowadays, many procedures that employ endophytic fungi as inoculants are being tested on *Aquilaria* and *Gyrinops*. It is conceivable that one day the inoculation technique will become more efficient, produces agarwood of super quality, and is cost effective.

References

Barden A, Awang Anak N, Mulliken T, Song M. Heart of the matter: agarwood use and trade and CITES implementation for *Aquilaria malaccensis*, TRAFFIC report, Malaysia; 2000.

CITES. Convention on International Trade in Endangered Species of Wild Fauna and Flora. Appendices I, II, and III of CITES. UNEP; 2005. 48 p.

Donovan DG, Puri RK. Learning from traditional knowledge of non-timber forest products: Penan Benalui and the autecology of *Aquilaria* in Indonesian Borneo. Ecol Soc. 2004;9(3):3.

Eurlings MCM, Heveling van Beek H, Gravendeel B. Polymorphic microsatellites for forensic identification of agarwood (*Aquilaria crassna*). Forensic Sci Int. 2010;197:30–4.

Gunn BV, Stevens P, Singadan M, Sunari L, Chatterton P. Eaglewood in Papua New Guinea. Resource management in Asia-Pacific working paper No. 51. Resource management in Asia-Pacific Program, Research School of Pacific and Asian Studies. The Australian National University, Canberra; 2004. 18 p.

Heuveling van Beek H, Phillips D. Agarwood: trade and CITES implementation in Southeast Asia. TRAFFIC report, Malaysia; 1999.

Jong PL, Tsan P, Mohamed R. Gas chromatography–mass spectrometry analysis of agarwood extracts from mature and juvenile *Aquilaria malaccensis*. Int J Agric Biol. 2014;16:644–8.

Kenmotsu Y, Asano K, Yamamura Y, Kurosaki F. Cloning and expression of putative Rac/Rop GTPase genes, *Am-rac1* and *Am-rac2*, involved in methyl jasmonate-induced transcriptional activation of farnesyl diphosphate synthase in cell cultures of *Aquilaria microcarpa*. Plant Mol Biol Rep. 2013;31:539–46.

Liu Y, Chen H, Yang Y, Zhang Z, Wie J, Meng H, Chen W, Feng J, Gan B, Chen X, Gao Z, Huang J, Chen B, Chen H. Whole-tree agarwood-inducing technique: an efficient novel technique for producing high-quality agarwood in cultivated *Aquilaria sinensis* trees. Molecules. 2013;18:3086–106.

Mohamed R, Jong PL, Zali MS. Fungal diversity in wounded stems of *Aquilaria malaccensis*. Fungal Div. 2010;43:67–74.

Mohamed R, Jong PL, Nurul Irdayu I. Succession patterns of fungi associated to wound-induced agarwood in wild *Aquilaria malaccensis* revealed from quantitative PCR assay. World J Microbiol Biotechnol. 2014;30(9):2427–36.

Mucharommah. Development of eaglewood (gaharu) in Bengkulu, Sumatera. In: Turjaman M, editor. Proceeding of gaharu workshop: development of gaharu production technology, a forest community based empowerment. Indonesia's Work Programme for ITTO PD425/06 Rev.1 (I); 2011. 134 p.

Ng LT, Chang YS, Kadir AA. A review on agar (gaharu) producing *Aquilaria* species. J Trop For Prod. 1997;2(2):272–85.

Paoli GD, Peart DR, Leighton M, Samsoedin I. An ecological and economic assessment of the non-timber forest product gaharu in the Gunung Palaung National Park, West Kalimantan, Indonesia. Conserv Biol. 2001;15:1721–32.

Pratiwi P, Santoso E, Turjaman M. Soil physical and chemistry of the gaharu (*Aquilaria* spp.) stands habitat in West Java. In: Turjaman M, editor. Proceeding of gaharu workshop: development of gaharu production technology, a forest community based empowerment. Indonesia's Work Programme for ITTO PD425/06 Rev.1 (I); 2011. 134 p.

Purnomo E, Turjaman M. The environmental characteristics of Kandangan site for gaharu plantation project. In: Turjaman M, editor. Proceeding of gaharu workshop: development of gaharu production technology, a forest community based empowerment. Indonesia's Work Programme for ITTO PD425/06 Rev.1 (I); 2011. 134 p.

Santoso E. Agarwood formation by fungal bioinduction technology. In: Susmianto A, Turjaman M, Setio P. Track record: agarwood inoculation technology by FORDA. FORDA Press, Bogor, Indonesia; 2013. 294 p.

Santoso E. Valuation of cultivated agarwood technology. FORDA Press, Bogor, Indonesia; 2015. 168 p.

Sitepu IR, Santoso E, Turjaman M. Fragrant wood gaharu: when the wild can no longer provide. Published by ITTO and FORDA, Bogor, Indonesia; 2011. 67 p.

Siran SA. Bioinduction agarwood: future elite commodity for exporting forestry sector. In: Susmianto A, Turjaman M, Setio P. Track record: agarwood inoculation technology by FORDA. FORDA Press, Bogor, Indonesia; 2013. 294 p.

Soehartono T, Newton AC. Conservation and sustainable use of tropical trees in the genus *Aquilaria* I. Status and distribution in Indonesia. Biol Conserv. 2000;96:83–94.

Soehartono T, Newton AC. Conservation and sustainable use of tropical tress in the genus *Aquilaria* II. The impact of gaharu harvesting in Indonesia. Biol Conserv. 2001a;97:29–41.

Soehartono T, Newton AC. Reproductive ecology of *Aquilaria* spp. in Indonesia. For Ecol Manage. 2001b;152:59–71.

Tamuli P, Boruah P, Nath SC, Leclercq P. Essential oil of eaglewood tree: a product of pathogenesis. J Essen Oil Res. 2005;17:601–4.

Tamuli P, Baruah P, Samanta R. Enzyme activities of agarwood (*Aquilaria malaccensis* Lamk.) stem under pathogenesis. J Spices Aromatic Crops. 2008;17(3):240–3.

Turjaman M, Santoso E, Tampubolon A. Progress research: agarwood cultivation and inoculation technology in Asia. In: Tampubolon AA et al., editors. National seminar: non-timber forest products. Mataram/Lombok: Forest Research Institute for NTFPs; 2012. p. 500.

Wong MT, Siah CH, Faridah QZ, Mohamed R. Characterization of wound responsive genes in *Aquilaria malaccensis*. J Plant Biochem Biotechnol. 2013;22(2):168–75.

Yang L, Qiao C, Xie D, Gong N-B, Lu Y, Zhang J, Dai J, Guo S. Antidepressant abietane diterpenoids from chinese eaglewood. J Nat Prod. 2013;76:216–22.

Chapter 5
Molecular Mechanism Studies of Terpenoid Biosynthesis in Agarwood

Zhi-Hui Gao and Jian-He Wei

Abstract The wound-induced natural production of agarwood hindered the development of its industry. Efficient and sustainable manner for its production should be developed. Metabolic engineering is one of the techniques with incredible potential to solve this problem. However, its realization should depend on the deep understanding of the whole biosynthetic pathway of the agarwood constituents. Here, we introduce studies of the molecular mechanism of the main agarwood constituents – terpenoids – to have an overview of the development in the study of the terpene biosynthesis pathway and the possible regulators of terpene biosynthesis in agarwood, including miRNAs and the members in the signal transduction. We also discuss the efforts that should be undertaken in the future.

5.1 Introduction

Agarwood is the resinous, fragrant wood produced by species of the tropical tree genus *Aquilaria*. It is in great demand for its high value in medicine, incense, and perfume across Asia (Yagura et al. 2005; Persoon and Heuveling van Beek 2008). However, agarwood is formed only after the trunk or branch is wounded or infected with microbes (Pojanagaroon and Kaewrak 2005; Bartel 2009). These stresses induce defense response and changes of secondary metabolism network, resulting in the accumulation of sesquiterpenes and phenylethyl-chromone derivatives and the deposition of resin in

Z.-H. Gao
Institute of Medicinal Plant Development, Chinese Academy of Medical Science & Peking Union Medicinal College, 151#Malianwabei Road, Beijing 100193, China

J.-H. Wei (✉)
Department of Culture Centre, Institute of Medicinal Plant Development, Chinese Academy of Medical Science & Peking Union Medicinal College, 151#Malianwabei Road, Beijing 100193, China

Hainan Branch, Institute of Medicinal Plant Development, Chinese Academy of Medical Science and Peking Union Medicinal College, 4#Yaogusi Road, Haikou 570311, China

Hainan Provincial Key Laboratory of Resources Conservation and Development of Southern Medicine, Wanning, 571533 China
e-mail: jhwei@implad.ac.cn

R. Mohamed (ed.), *Agarwood*, Tropical Forestry,
DOI 10.1007/978-981-10-0833-7_5

the tree (Ito and Kumeta 2010). Thus, natural production of agarwood is extremely limited. Due to overexploitation, the population of the resource genus *Aquilaria* is dramatically declining, and nine species have been listed in the IUCN Red List of Threatened Plants published by The World Conservation Union (IUCN) since 1998 (Oldfield et al. 1998). Furthermore, *Aquilaria* species are regulated under the Convention on International Trade in Endangered Species of Wild Fauna and Flora (CITES).

On the other hand, even though *Aquilaria* trees have been planted in South China and many Southeast Asian countries on a large scale, the global demand for agarwood still considerably exceeds the supply (Gao et al. 2014). The major restriction of the agarwood production is caused by the limited knowledge of the mechanism of the agarwood induction. Further understanding of the regulation and molecular mechanism of agarwood formation will help to reveal the relationship between plant stress response and secondary metabolism and supply information for efficient agarwood production without damage to the wild resource of *Aquilaria*.

Since sesquiterpenes and phenylethyl-chromone derivatives are the main compounds in agarwood (Ishihara et al. 1993; Yagura et al. 2003; Ito and Kumeta 2010; Chen et al. 2011, 2012), understanding the biosynthesis and regulation of sesquiterpenes and chromone in *Aquilaria* spp. is critically important in determining the mechanism of agarwood formation. Sesquiterpenes belong to a large secondary metabolite family – terpenoids. A recent study has reported that the sesquiterpene β-caryophyllene from the essential oil of *Aquilaria crassna* demonstrated selective antiproliferative effects against colorectal cancer, as well as the antimicrobial and antioxidant properties (Dahham et al. 2015). This result underlined the great potential medicinal value of sesquiterpenes from agarwood. In the plant kingdom, the terpenoid biosynthesis pathway has been well studied. As shown in Fig. 5.1, all terpenoids are derived from the common precursor isopentenyl diphosphate (IPP), which is synthesized via two different pathways, the mevalonic acid (MVA) and methylerythritol phosphate (MEP) pathways (Tholl 2006). Then, the IPPs were adopted to the condensation reactions to generate the linear polymers. In the final step, terpenes are formed by terpene synthases (TPS) from the linear polymers.

Studies of the induction mechanism of the terpenoid biosynthesis of agarwood showed that the methyl jasmonate (MJ) treatment increases the expression of TPS genes and induces the production of sesquiterpene δ-guaiene in cell cultures of *A. crassna* and *Aquilaria sinensis* (Ito and Kumeta 2010; Xu et al. 2013a). We also found that the pruning of actively growing saplings of *A. sinensis* resulted in hydrogen peroxide (H_2O_2) burst, and exogenous H_2O_2 could induced *A. sinensis* plants to form vessel occlusions and produce sesquiterpenes (Zhang et al. 2014b). Further study in cultured cell suspensions of *A. sinensis* found that H_2O_2 could promote programmed cell death (PCD), salicylic acid (SA) accumulation, and sesquiterpene production (Liu et al. 2015a). On the other hand, by pyrosequencing the small rRNA subunits of the non-contaminated wound sample and sample opened to the environment, no substantial variation was observed either in fungal and bacterial enrichment and diversity or in the relative abundances of taxa (Zhang et al. 2014a). This demonstrated that wound in the absence of variations in microbial communities may induce agarwood formation. This result does not support the long-standing notion that agarwood formation depends on microbes.

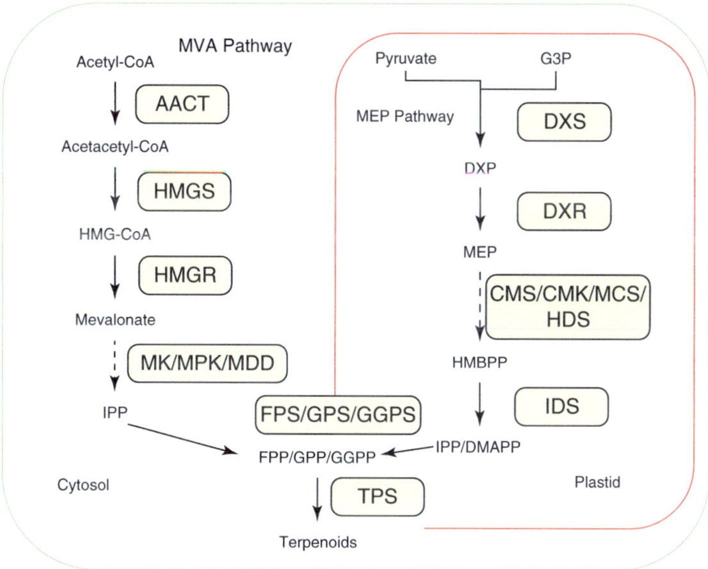

Fig. 5.1 Biosynthetic pathways and their compartmentalization leading to terpenoids in plants. *AACT* acetyl-CoA C-acetyl transferase, *CMK* 4-(cytidine 50-diphospho)-2-C-methyl-D-erythritol kinase, *DMAPP* dimethylallyl diphosphate, *DXP* 1-deoxy-D-xylulose 5-phosphate, *DXR* DXP reductoisomerase, *DXS* DXP synthase, *FPS* farnesyl diphosphate synthase, *FPP* farnesyl diphosphate, *GPS* geranyl diphosphate synthase, *GGPS* geranylgeranyl diphosphate synthase, *GGPP* geranylgeranyl diphosphate, *GPP* geranyl diphosphate, *HMBPP* (E)-4-hydroxy-3-methylbut-2-enyl diphosphate, *HMG-CoA* 3-hydroxy-3-methylglutaryl-CoA, *HMGR* HMG-CoA reductase, *HMGS* HMG-CoA synthase, *IPP* isopentenyl diphosphate, *CMS* 2-C-methyl-erythritol-4-P cytidyl transferase, *CMK* 4-(cytidine-59-diphospho)-2-C-methyl-D-erythritol kinase, *MCS* 2-C-methyl-D-erythritol-2,4-cyclo diphosphate synthase, *HDS* (E)-4-hydroxy-3-methylbut-2-enyl diphosphate synthase, *IDS isopentenyl* diphosphate synthase, *MEP* 2-C-methyl-D-erythritol 4-phosphate, *MK* mevalonate kinase, *MPK* phosphomevalonate kinase, *MDD* mevalonate diphosphate decarboxylase, *TPS* terpene synthase. Enzyme names are *boxed*

In recent years, several groups have focused on the cloning and enzyme identification of the sesquiterpene biosynthesis genes and regulatory genes in *Aquilaria* spp… In this chapter, we present an overview on the development of these studies in elucidating biosynthesis mechanism of terpene-based compounds in agarwood.

5.2 Terpene Synthases in *Aquilaria* Species

5.2.1 Enzymes in Sesquiterpene Synthesis

The first report of the terpene synthase in *Aquilaria* species is the δ-guaiene synthases from *A. crassna* (Ito and Kumeta 2010). It was found that the production of sesquiterpene δ-guaiene was induced, and the TPS might be selectively activated in the methyl jasmonate (MJ)-treated cells of *A. crassna*. Thus, to isolate the genes responsible for the sesquiterpene synthesis, a cDNA library from RNA isolated from MJ-treated cells of *A. crassna* was constructed. Five clones with very similar amino acid sequences were isolated by screen using PCR methodologies. These clones were expressed in *Escherichia coli*, and enzymatic reactions using farnesyl pyrophosphate revealed that three of the clones yielded the same compounds as extracted from MJ-treated cells, the major product being δ-guaiene. These genes and their encoded enzymes are the first sesquiterpene synthases yielding guaiene-type sesquiterpenes as their major products to be reported. Expression of the fourth terpene synthase gene in bacteria resulted in the accumulation of the protein in insoluble forms. Site-directed mutagenesis of the inactive clone and three-dimensional homology modeling suggested that the structure of the N-terminal domain was important in facilitating proper folding of the protein to form a catalytically active structure.

To identify the primary genes that may be related to agarwood formation, our group used the high-throughput technique to study the transcriptome of *A. sinensis* (Xu et al. 2013a). Two cDNA libraries from the healthy and wounded stems (designated H and W) were constructed using switching mechanism at 5′ end of RNA transcript (SMART) technology. The two libraries were sequenced in one run using the 454 GS-FLX pyrosequencing technology. After de novo assembly of these reads, 67,972 unigenes from the H library and 74,139 unigenes from the W library were generated, respectively.

Based on the degenerated primers, full-length cDNAs of three sesquiterpenes synthases (*ASS1*, *ASS2*, and *ASS3*) were cloned. All of them have open reading frames (ORFs) of 1644 nucleotides, which encode almost the same protein of 547 amino acids. Their nucleotide and protein sequences have more than 92 % identity with one another. These proteins contain two motifs (the RRx_8W motif at the N-terminus and the DDxxD motif known to be a divalent metal ion substrate-binding site) that are functionally important and highly conserved in all terpene synthase proteins.

To confirm that *ASS1-3* encode active ASSs, they were cloned into the pET-28a vector and heterologously expressed in *E. coli*. The enzyme assays were performed using the geranyl diphosphate (GPP) (C10), farnesyl diphosphate (FPP) (C15), and geranylgeranyl diphosphate (GGPP) (C20) substrates, and the reaction products were analyzed by GC-MS. The result showed that ASS1-3 did not accept GPP or GGPP as substrate and converted only FPP to terpene products. Furthermore, the three enzymes yielded the same compounds; the major product was identified as δ-guaiene (74.2 %), and the minor products as β-elemene (16.3 %) and α-guaiene (9.5 %).

The expression profile of *ASS* was analyzed using reverse transcription-PCR (RT-PCR). MJ-treated calli were also used in this investigation. The results demonstrated that *ASS1* and *ASS2* were both induced by either mechanical wounding or MJ treatment. The expression of *ASS1* was upregulated significantly – approximately 800 times in response to mechanical wound and 1000 times in response to MJ treatment after 5 days.

5.2.2 Enzymes in Triterpene Synthesis

In a more recent research, in vitro materials from *Aquilaria agallocha* were found to contain abundant amounts of triterpene cucurbitacin I and E (Chen et al. 2014). Although the true identity of the *A. agallocha* species is dubious as it is a synonym of *Aquilaria malaccensis*, nevertheless the assembled draft of its genome contributes to sequence information on *Aquilaria*, which is currently lacking. To investigate the cucurbitacin pathway in *A. agallocha*, transcripts were annotated using *Arabidopsis thaliana* proteins as well as UniProt enzymes. The annotated transcripts were then used to infer a putative cucurbitacin pathway in *A. agallocha* by referring to the mevalonate pathway in *A. thaliana* from KEGG (Kanehisa et al. 2004). The identified genes, which encode for enzymes in the cucurbitacin E and I pathway, include three important gene categories: 3-hydroxy-3-methylglutaryl-coenzyme A reductase (HMGR), 1-deoxy-d-xylulose-5-phosphate synthases (DXS), and squalene synthetase (SQS), as well as genes which encode for synthases in cucurbitacin metabolism.

The putative pathway also includes many differentially expressed cytochrome P450s (CYP450s) and S-adenosyl-L-methionine-dependent methyltransferases (SAM-Mtases) as putative downstream genes. CYP450s are one of the largest gene families in plants and catalyze most oxidation steps in secondary metabolism such as in the biosynthesis of defense compounds, pigment, and antioxidants (Bak et al. 2011; Rasool and Mohamed 2015). Putatively, CYP450s may catalyze the conversion of cucurbitadienol. SAM-Mtases may also act on cucurbitadienol by catalyzing methylation, as it is known that many compounds with anti-microorganism functions have cucurbitadienol backbones activated by methylation (Struck et al. 2012). This research annotated 161 cytochrome P450s and 66 SAM-Mtases in the *A. agallocha* genome, of which, 66 CYP450s and 27 SAM-Mtases showed significant up-regulation by MJ treatment. These genes can be considered candidate genes that are possibly involved in the cucurbitacin pathway. As well, they identified a small number of SAM-Mtases that contained the VOZ cis-element, though their expression was not observed to be significantly up-regulated.

5.3 Isoprene Biosynthetic Pathway in *Aquilaria sinensis*

In our study of *A. sinensis* transcriptome, 30 unigenes were annotated as being related to the sesquiterpene biosynthesis pathway, including three HMGRs, twelve DXSs, two farnesyl diphosphate synthases (FPSs), and five sesquiterpene synthases. Unigenes putatively encoding other enzymes in the sesquiterpene biosynthesis pathway were also found, including acetyl-CoA C-acetyl transferase (AACT), mevalonate kinase (MK), phosphomevalonate kinase (MPK), and 2-C-methyl-D-erythritol-2,4-cyclodiphosphate synthase (MCS).

FPS is a branch-point enzyme in terpenoid biosynthesis (Matsushita et al. 1996; Cao et al. 2012). Its product, FPP, is the common precursor of sesquiterpenes, steroids, and farnesylated proteins. In our transcriptome libraries, two unigenes that putatively encode FPS were identified. One is a full-length cDNA (named *AsFPS*) that has an open reading frame (ORF) of 1029 nucleotides and encodes a protein of 342 amino acids. AsFPS had the highest homology with the FPS sequence from *A. microcarpa*, suggesting that it belongs to the FPS family. With the exception of this AsFPS, the full-length ORFs of other unigenes were unavailable. However, based on the transcriptome sequencing, more genes have been cloned and characterized later.

5.3.1 *Cloning and Characterization of* AsDXS

DXS is the first rate-limiting enzyme for sesquiterpene synthesis in the MEP pathway. Using 5′ and 3′ rapid amplification of cDNA ends (RACE), three cDNAs of *DXS* genes were cloned and characterized from *A. sinensis* (Xu et al. 2014). The total length of ORF and UTR segments for *AsDXS1*was 2374 bp, with a 56-bp 5′-UTR, 203-bp 3′-UTR, and 2154-bp ORF encoding a protein of 717 amino acids. The total length of ORF and UTR segments for *AsDXS2* was 2257 bp, with a 118-bp 3′-UTR, poly(A) tail of 33 bp, and 2139-bp ORF encoding a protein of 712 amino acids. The full-length cDNA of *AsDXS3* was 2761 bp, with a 280-bp 5′-UTR, 342-bp 3′-UTR, and 2139-bp ORF encoding a protein of 712 amino acids. Their proteins have similarity of between 51.74 and 67.92 %, with *AsDXS3* being the most divergent.

These genes represent phylogenetically distinct clades conserved among plants. Functional complementation in a DXS-deficient *E. coli* strain EcAB4-2 demonstrated that they are active DXS, which rescued the *E. coli* mutant. Their expression profiles in different tissues and in response to different treatments were analyzed by RT-PCR. All three genes are highly expressed in the stem, followed by the leaf and root. *AsDXS1* was significantly stimulated by mechanical, chemical, and H_2O_2 treatment, whereas *AsDXS2* and *AsDXS3* only responded to chemical treatment and mechanical treatment, respectively. All three genes were fluctuating in response to MJ treatment, with expression peaks occurring at different time points. Our results suggested the conservation of DXS in evolution and imply their distinct functions in primary and defensive sesquiterpene metabolism in *A. sinensis*.

Further characterization of the function of DXS genes in *A. sinensis* necessitates the creation of a transgenic plant and detection of whether DXS is directly related to agarwood sesquiterpene formation. This study will facilitate further investigation of AsDXS physiological functions and their regulatory roles in sesquiterpene biosynthesis and thus in agarwood formation.

5.3.2 *Cloning and Regulation of* AsHMGS

Hydroxylmethylglutaryl-CoA synthase (HMGS) is a key enzyme in the MVA pathway. Homologous HMGS gene cDNA was isolated from the stem of *A. sinensis* through RT-PCR and RACE and named as *AsHMGS* (Liu et al. 2014a). The full length of the *AsHMGS* gene was 1831 bp containing a 1398 bp open reading frame, a 157 bp 5′-UTR, and a 276 bp 3′-UTR. The predicted protein has 465 amino acids and the molecular weight is 51.4 kD. Representative motifs and active site were deduced in the amino acid sequence of *AsHMGS*. Sequence comparing and phylogenetic analysis suggested that *AsHMGS* had high similarity to *HMGS* genes from *Arabidopsis thaliana*, *Arabidopsis lyrata* subsp. *lyrata*, and *Brassica juncea*, followed by *Camellia sinensis*, *Camptotheca acuminate*, and *Panax ginseng*. Gene expression analysis by RT-PCR showed that in calli of *A. sinensis*, the transcription of *AsHMGS* could be greatly induced by MJ, and the express pattern after MJ treatment was highly similar with *AsHMGR*. This indicated that these two genes might be co-regulated by same regulators.

5.3.3 Cloning and Regulation of AsHMGRs

HMGR is the first rate-limiting enzyme for sesquiterpene synthesis in the MVA pathway. RT-PCR and RACE were used to clone the full-length cDNA of HMGR from *A. sinensis* based on the conserved *HMGR* gene fragments (Xu et al. 2013b). The cDNA sequence of *HMGR* was obtained from *A. sinensis* by taking the advantage of the conserved domains of previously cloned *HMGRs*. Using newly designed degenerate primers corresponding to two conversed domains found in most HMGRs (–ILGQCCE-, and-SHMKYNR-), a 1047 bp conserved fragment was obtained. Then the 5′ and 3′ RACE were carried out, and a full-length cDNA (named *AsHMGR1*) putatively encoding an HMGR was deduced and confirmed by sequencing. The full-length cDNA was 2026 bp, with a 135 bp 5′-UTR, a 352 bp 3′-UTR, a poly(A) tail of 29 bp, and a 1719 bp ORF encoding a protein of 572 amino acids. The predicted AsHMGR1 protein had a calculated molecular weight of 6.17 × 104 and a pI value of 6.24 (http://www.cn.expasy.org/tools/protparam.html).

TMHMM2.0 analysis predicted that AsHMGR1 contains two transmembrane domains, one located between F39 and A61 and the other between M81 and I103. Multiple sequence alignment demonstrated that AsHMGR1 had high amino acid sequence similarity to *Vitis vinifera* (78.34 %), *Tilia miqueliana* (78.27 %), *Corylus avellana* (76.70 %), and *M. alba* (75.92 %), suggesting that AsHMGR1 belongs to the HMGR superfamily. The *N*-terminal end of HMGR was significantly diverse in both length and composition, while the *C*-terminal catalytic domain was highly conserved. The putative amino acid sequence of AsHMGR1 contains two HMG-CoA-binding motifs ("EMPIGY-VQIP" and "TTEGCLIA") and two NADP(H)-binding motifs ("DAMGMNM" and "GTVGGGT"). These motifs are functionally important and highly conserved in all HMGR proteins. A phylogenetic tree was constructed using the Clustal W program based on *HMGR* sequences obtained from GenBank. This revealed that *AsHMGR1* had the highest homology with the HMGR (ABK88909) cloned from *Atractylodes lancea*.

Expression analysis by RT-PCR showed that *AsHMGR1* expression level was highest in the stems, followed by roots and branches, and lowest in the leaves. We treated *A. sinensis* calli with MJ and H_2O_2 and examined the expression of *AsHMGR1* after treated for 2, 4, 6, 8, 12, and 24 h. The results showed that *AsHMGR1* was gradually induced and reached the maximum at 12 h when treated with MJ, while treated with H_2O_2, expression of *AsHMGR1* was induced from 2 to 8 h and reached the maximum at 6 h and then decreased immediately.

Another HMGR was cloned with the specific primers designed according to the transcript sequence from the *A. sinensis* transcriptome database (Xu et al. 2013c) by RT-PCR and 5′ rapid RACE technology. The length of *AsHMGR2* ORF was 1749 bp, encoding 582 amino acids. Motif prediction showed that the AsHMGR2 have two motifs combining the HMG-CoA that were EMPVGYVQIP and TTEGCLVA. The Glu in the later motif is critical for the catalytic activity. The AsHMGR2 also have two NADPH-combining motifs. Tissue expression analysis indicated that *AsHMGR2* was mainly expressed in roots and shoot tips, followed by the stems, and was lowest in the leaves. In addition, this gene could be induced by mechanical wound as well as chemical liquid induction and reached the highest expression level at 6 h and 8 h, respectively. These studies will provide a foundation for further research on its function in agarwood sesquiterpene biosynthesis.

5.3.4 Cloning and Regulation of AsAACT

The AACT is a key enzyme in the early steps of terpenoid biosynthesis. One unique sequence containing partly *AACT* gene sequence was discovered in our previous transcriptome dataset of *A. sinensis*. The *AsAACT* gene was also cloned by RT-PCR and RACE strategy with the template of RNA extracted from *A. sinensis* stem (Liu et al. 2014b). The full length of *AsAACT* cDNA was 1476 bp, containing a 1236 bp ORF that encoded 411 amino acids. The bioinformatic analysis showed that the enzyme was predicted to be located in the cytoplasm. The *AsAACT* expression in calli was analyzed with GADPH gene as an internal control gene in wounded condition by qRT-PCR technique. The results showed that the *AsAACT* expression could be induced, and the highest expression level in calli was obtained at 4 h after the MJ treatment.

5.3.5 Cloning and Regulation of FPS *in* A. sinensis *and* Aquilaria microcarpa

FPS is one of the key rate-limiting enzymes in the sesquiterpene metabolic pathways. A cDNA clone, designated *Am-FaPS-1*(1310 bp), was isolated from callus culture of *A. microcarpa* by RACE (Kenmotsu et al. 2011). This gene contains an ORF encoding the protein of 342 amino acid residues with high homology to FPS from various plant sources. Two characteristic motifs for chain length determination (I) and active site lid (II), the common structures of FPS, are well conserved, and the *Am-FaPS-1* product also contains a substrate-Mg^{2+}-binding site as the unique structure of FaPSs. An appreciable increase in the transcriptional level of *Am-FaPS-1* was observed by the exposure of the cell culture to MJ, yeast extract, and Ca^{2+}-ionophore A23187. These results suggested that *Am-FaPS-1* and its translate play roles in MJ- and yeast extract-induced responses of *A. microcarpa* and Ca^{2+} functions as an important messenger molecule in these processes.

The ORF of FPS was cloned by PCR based on the transcript sequence of *AsFPS1* from the *A. sinensis* transcriptome database and sequenced. The result showed that the full length of *AsFPS1* was 1342 bp with the ORF 1029 bp, encoding 342 amino acids. Bioinformatic analysis showed that the amino acid sequence has over 86% identity with FPSs from nine other plant species and contains two highly conserved motifs DDXXD.

Total RNA was extracted from the roots, stems, and leaves of 3-year-old *A. sinensis*, and from healthy and wounded *A. sinensis* calli, to detect the expression pattern of *AsFPS1* in different tissues and its expression profile in responding to different treatments (Yang et al. 2013). Tissue expression analysis indicated that *AsFPS1* was mainly expressed in roots and stems and was lower in leaves. The gene was induced by mechanical wound as well as chemical liquid induction and reached the highest expression level at 6 h and 12 h, respectively. The full-length cDNA clone of *AsFPS1* and its expression patterns analysis will provide a foundation for the following study on its biological function and agarwood sesquiterpene biosynthesis mechanism.

5.3.6 Cloning and Regulation of AsDXR

1-Deoxy-D-xylulose 5-phosphate reductoisomerase (DXR) is the second key enzyme in the plastid MEP pathway of sesquiterpene biosynthesis. A DXR gene was cloned from the stem of *A. sinensis* by the methods of RT-PCR and RACE technique and named as *AsDXR* (Liu et al. 2015b). The full-length cDNA of *AsDXR* was 1768 bp, containing a 1437 bp ORF encoding a polypeptide of 478 amino acids with a molecular weight of 51.859 kD and the theoretical isoelectric point of 6.29. Comparative and bioinformatic analysis of the deduced AsDXR protein showed extensive homology with DXRs from other plant species, especially *Theobroma cacao* and *Gossypium barbadense*, and contained a conserved transit peptide for plastids and extended pro-rich region and a highly conserved NADPH-binding motif owned by all plant DXRs. Southern blot analysis indicated that *AsDXR* belonged to a small gene family. Tissue expression pattern analysis revealed that *AsDXR* expressed strongly in the root and stem, but weakly in the leaf. Additionally, *AsDXR* expression was found to be activated by MJ especially at the 12 h time point after treatment (34-fold increasement). On the other hand, the contents of three sesquiterpenes (δ-guaiene, β-humulene, and α-guaiene) were significantly induced by MJ. This study enables us to further elucidate the role of AsDXR in the biosynthesis of agarwood sesquiterpenes in *A. sinensis* at the molecular level.

5.4 Regulator Genes in Agarwood Production

5.4.1 Identification of MicroRNAs in A. sinensis

MicroRNAs (miRNAs) are key gene expression regulators involved in various plant stress response and metabolic processes, with possible role in agarwood formation. However, no report concerning miRNAs in *Aquilaria* is available. The mature miRNAs regulate gene expression mainly by targeting mRNA for mRNA cleavage or repressing the protein translation (Bartel 2009). In plants, the miRNA and the target mRNA interact in a near-perfect sequence complementary way (Bartel and Bartel 2003). Thus, the most direct function study of miRNA is conducted by predicting the target gene. In the effort to find out miRNAs related to agarwood production, the small RNA high-throughput sequencing and 454 transcriptome data were adopted to identify both conserved and novel miRNAs in *A. sinensis* (Gao et al. 2012, 2014). Their expressions in the stems were monitored at 0.5, 4, and 48 h after wounding treatment. The miR171, miR390, miR394, miR2111, and miR3954 families remained at the reduced level two days after the treatment. A total of 131 homologous miRNAs in the 0.5 h library showed over threefold variation of read number compared to the control library, of which 12 exhibiting strong expression alterations were further confirmed by RT-PCR. Among them, miR396a2, miR164a2, miR396b2 and miR159a, miR156a, miR171a, miR171b, miR168b, miR394, and miR3954 were downregulated by the wound treatment, while miR172 and miR169 were upregulated. Target prediction and annotation of the miRNAs demonstrated that the binding, metabolic process, catalytic activity, and cellular process are the most common functions of the predicted targets of these newly identified miRNAs in *A. sinensis*. The targets of miR396-4 and miR394-1 were predicted having aspartic-type endopeptidase activity, which might also act as a part of defense response. For the candidate novel miRNAs, a possible target of novel-100b-5p was predicted to have glutathione peroxidase activity and response to oxidative stress, which always happened after wound. These wound-related miRNAs may have important functions in agarwood formation; however, because of limited genomic information of *A. sinensis*, targets that directly relate to agarwood biosynthesis, specifically sesquiterpene synthesis, were not identified. Moreover, many of the wound-related miRNAs had not identified any targets. Since most miRNA targets are transcript factors, this work provides reference for future study on transcript factors.

5.4.2 Identification of the Regulator Genes in A. microcarpa

To study the signal transduction of the MJ-induced sesquiterpene biosynthesis, two cDNA clones presumably encoding calmodulin (CAM), a Ca^{2+}-binding protein, and two cDNA clones presumably encoding Rac/Rop GTPases, a plant-specific monomeric GTP-binding protein, were isolated based on the reported amino acids sequences from *A. microcarpa* calli (Kenmotsu et al. 2010, 2013). These clones were designated *Am-cam-1*, *Am-cam-2*, *Am-rac1*, and *Am-rac2*, respectively. The *Am-cam-1* and *Am-cam-2* were all 450 bp and their nucleotide sequences were very similar, and only six nucleotides were found to be replaced. The putative amino acid sequences of the translational products of the two genes showed high homology with those of CAMs from various plant sources. They contained four highly conserved Asp-rich motifs characteristic of proteins that correspond to Ca^{2+}-binding domains. Furthermore, the proteins with the expected weight (approximately 17 kDa) could be obtained by expressing them in *E. coli*, strongly suggesting that both *Am-cam-1* and *Am-cam-2* genes encode CAM proteins in *A. microcarpa*.

The transcription of *Am-cam1* and *Am-cam-2* was stimulated by the treatment with MJ and yeast extract. In contrast, Ca^{2+}-ionophore A23187 did not show inducing activity for the expression of these two calmodulin genes. Transcriptional levels of *Am-rac1* remained steady, while those of *Am-rac2* increased dramatically following treatment of cultured cells with either a yeast extract or MJ. Moreover, *Am-FaPS1*, which was transcriptionally activated in *A. microcarpa* cells grown in the presence of MJ, showed markedly lowered expression levels in the presence of various signal transduction-related inhibitors involved in Ca^{2+}-, Rap/Rop GTPase-, or ubiquitin-dependent signaling processes, whereas the expression of *Am-FaPS1* was markedly increased, even in the absence of MJ in *A. microcarpa* cells overexpressing *Am-rac2*. These findings suggested that Rac/Rop GTPase proteins played important roles in jasmonate-induced enhancement of terpenoid metabolism in *A. microcarpa*. Studies using the specific inhibitors for calmodulin- and Rac GTPase-dependent cellular processes showed that the enhancement of the expression of δ-guaiene synthase (δ-GS), a sesquiterpene biosynthetic enzyme gene in *A. microcarpa*, in response to MJ stimulation was markedly reduced by the inhibitor treatment in a dose-dependent manner. These observations suggested that calmodulin- and Rac GTPase-dependent cellular processes were involved in transcriptional activation of δ-GS in *A. microcarpa* cells triggered by MJ.

Transformation of *A. microcarpa* cell culture by *cam1* and *rac2* triggered the transcriptional activation of δ-GS even in the absence of MJ (Kurosaki and Taura 2015). In contrast, *A. microcarpa* cultures overexpressing mutated *rac2* of which translate is lacking in binding ability toward GTP did not show the significant expression of δ-GS gene. These results suggested that calmodulin and Rac GTPase proteins function as the important mediators in jasmonate-induced activation of sesquiterpene biosynthesis in *A. microcarpa*.

According to studies of the regulation mechanism of terpene biosynthesis in agarwood, a hypothetical regulation scheme was proposed (Fig. 5.2). This scheme concludes the function and possible relation of MJ, H_2O_2, ethylene (ET), Ca^{2+}, miRNA, SA, and PCD in the wound-induced defensive responses and terpene biosynthesis. A large amount of information on the molecular mechanism is yet to be gathered in each of the signal pathway, as well as their relationships. More protein regulators and other regulators such as miRNAs and their functions are waiting to be identified.

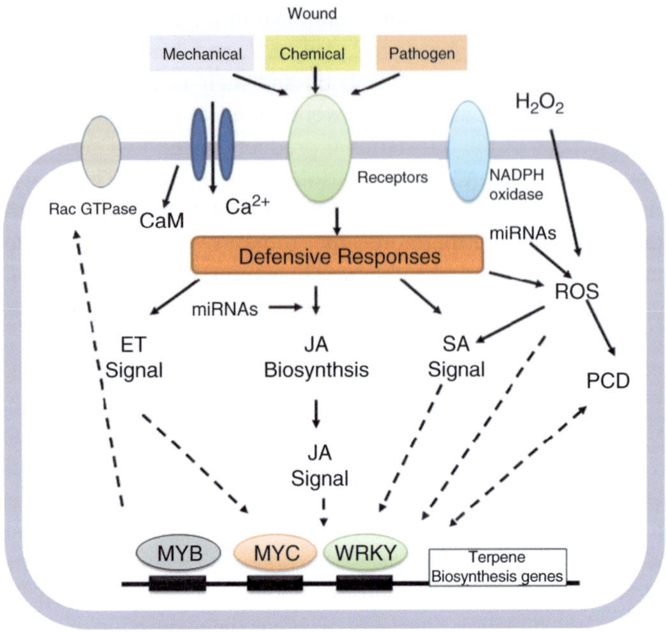

Fig. 5.2 Proposed mechanism of wound-induced terpene biosynthesis and regulation. Wound signals contacted with their receptors, stimulate the Ca^{2+} pathway, and induce the defensive responses downstream, including the jasmonic acid (*JA*) signals, the hydrogen peroxide (H_2O_2) pathway, the salicylic acid (*SA*) signals, and the ethylene (*ET*) signals. miRNAs may have roles in controlling the JA biosynthesis and the H_2O_2 pathways. Reactive oxygen species (*ROS*) in the H_2O_2 pathway may induce the SA signal and the programmed cell death (*PCD*). These defensive signals then activate the transcription factors (MYB, MYC, and WRKY) downstream. The activated transcription factor binds to the cis-acting element in the promoter of terpene biosynthesis genes and starts their transcription

5.5 Conclusion and Future Perspective

As a high-priced non-timber product with values in culture, medicine, as well as beauty industry, attraction on agarwood keeps on increasing from all over the world. Many techniques from seedling cultivation to harvesting and processing have been developed, but the most critical – agarwood induction for commercial production – is still inefficient. A more efficient and sustainable method for high-quality product is highly desirable and biological synthesis could be one approach. Perhaps we could learn from the experience of artemisinin, a sesquiterpenoid pharmaceutical, which semisynthetic production is the first success story for the combined use of metabolic engineering and synthetic biology at industrial scale. Its success had very much depended on deep understanding of the whole biosynthesis pathway (Paddon and Keasling 2014). Therefore, efforts to identify the enzymes in the terpenoid synthesis pathway of agarwood should be the fundamental work if we were to follow this example. In our previous description, only limited enzymes have been identified or cloned from *Aquilaria* spp. A large number of enzymes in the common upstream pathway of terpenoid synthesis and their functions are still in need to be clarified. On the other hand, identification of terpene synthases and CYP450s in the downstream pathway would be a difficult task because both of them have big families. Analyzing these enzymes at this scale should be facilitated by modern techniques such as genomics and transcriptomics. Also it needs to be pointed out that the production of high-quality agarwood requires the detailed knowledge of the agarwood constituents and their medicinal roles. This knowledge will guide us to focus on important constituents and lead to the level at which a specific constituent of interest is being produced in the synthesis pathway.

References

Bak S, Beisson F, Bishop G, Hamberger B, Hofer R, Paquette S, Werck-Reichhart D. Cytochromes p450. Arabidopsis Book. 2011;9:e0144.

Bartel DP. MicroRNAs: target recognition and regulatory functions. Cell. 2009;136(2):215–33.

Bartel B, Bartel DP. MicroRNAs: at the root of plant development? Plant Physiol. 2003;132(2): 709–17.

Cao X, Yin T, Miao Q, Li C, Ju X, Sun Y, Jiang J. Molecular characterization and expression analysis of a gene encoding for farnesyl diphosphate synthase from *Euphorbia pekinensis* Rupr. Mol Biol Rep. 2012;39(2):1487–92.

Chen H, Yang Y, Xue J, Wei J, Zhang Z. Comparison of compositions and antimicrobial activities of essential oils from chemically stimulated agarwood, wild agarwood and healthy *Aquilaria sinensis* (Lour.) Gilg trees. Molecules. 2011;16(6):4884–96.

Chen HQ, Wei JH, Yang JS, Zhang Z, Yang Y, Gao ZH, Sui C, Gong B. Chemical constituents of agarwood originating from the endemic genus *Aquilaria* plants. Chem Biodivers. 2012;9(2): 236–50.

Chen CH, Kuo TC, Yang MH, Chien TY, Chu MJ, Huang LC, Chen CY, Lo HF, Jeng ST, Chen LF. Identification of cucurbitacins and assembly of a draft genome for *Aquilaria agallocha*. BMC Genomics. 2014;15(1):578.

Dahham SS, Tabana YM, Iqbal MA, Ahamed MB, Ezzat MO, Majid AS, Majid AM. The anticancer, antioxidant and antimicrobial properties of the sesquiterpene beta-caryophyllene from the essential oil of *Aquilaria crassna*. Molecules. 2015;20(7):11808–29.

Gao ZH, Wei JH, Yang Y, Zhang Z, Xiong HY, Zhao WT. Identification of conserved and novel microRNAs in *Aquilaria sinensis* based on small RNA sequencing and transcriptome sequence data. Gene. 2012;505(1):167–75.

Gao ZH, Yang Y, Zhang Z, Zhao WT, Meng H, Jin Y, Huang JQ, Xu YH, Zhao LZ, Liu J, Wei JH. Profiling of microRNAs under wound treatment in *Aquilaria sinensis* to identify possible microRNAs involved in agarwood formation. Int J Biol Sci. 2014;10(5):500–10.

Ishihara M, Tsuneya T, Uneyama K. Sesquiterpene constituents in agarwood. 4. Fragrant sesquiterpenes from agarwood. Phytochemistry. 1993;33(5):1147–55.

Ito M, Kumeta Y. Characterization of delta-guaiene synthases from cultured cells of *Aquilaria*, responsible for the formation of the sesquiterpenes in agarwood. Plant Physiol. 2010;154(4): 1998–2007.

Kanehisa M, Goto S, Kawashima S, Okuno Y, Hattori M. The KEGG resource for deciphering the genome. Nucleic Acids Res. 2004;32(Database issue):D277–80.

Kenmotsu Y, Yamamura Y, Ogita S, Katoh Y, Kurosaki F. Transcriptional activation of putative calmodulin genes *Am-cam-1* and *Am-cam-2* from *Aquilaria microcarpa*, in response to external stimuli. Biol Pharm Bull. 2010;33(11):1911–4.

Kenmotsu Y, Ogita S, Katoh Y, Yamamura Y, Takao Y, Tatsuo Y, Fujino H, Kadota S, Kurosaki F. Methyl jasmonate-induced enhancement of expression activity of *Am-FaPS-1*, a putative farnesyl diphosphate synthase gene from *Aquilaria microcarpa*. J Nat Med. 2011;65(1):194–7.

Kenmotsu Y, Asano K, Yamamura Y, Kurosaki F. Cloning and expression of putative Rac/Rop GTPase genes, *Am-rac1* and *Am-rac2*, involved in methyl jasmonate-induced transcriptional activation of farnesyl diphosphate synthase in cell cultures of *Aquilaria microcarpa*. Plant Mol Biol Rep. 2013;31:539–46.

Kurosaki F, Taura F. Transcriptional activation of sesquiterpene biosynthetic enzyme δ-guaiene synthase gene in cell cultures of *Aquilaria microcarpa* overexpressing *cam1* and *rac2* encoding calmodulin and Rac GTPase. Plant Gene. 2015;2:25–8.

Liu J, Xu Y, Yang Y, Liang L, Gao Z, Yang Y, Zhang Z, Sui C, Wei J. Cloning and gene expression of 3-Hydroxy-3-Methylglutaryl-CoA synthase gene (*AsHMGS*) from *Aquilaria sinensis* (Lour.) Gilg. Bulletin of Botanical Research. 2014a;34(1):75–84.

Liu J, Xu YH, Yang Y, Liang L, Han XM, Gao ZH, Zhang Z, Wei JH. Cloning and gene expression of acetyl-CoA C-acetyl transferase gene (*AsAACT*) from *Aquilaria sinensis*. Zhongguo Zhong Yao Za Zhi. 2014b;39(6):972–80.

Liu J, Xu Y, Zhang Z, Wei J. Hydrogen peroxide promotes programmed cell death and salicylic acid accumulation during the induced production of sesquiterpenes in cultured cell suspensions of *Aquilaria sinensis*. Funct Plant Biol. 2015a;42(4):337–46.

Liu J, Xu Y, Liang L, Wei J. Molecular cloning, characterization and expression analysis of the gene encoding 1-deoxy-D-xylulose 5-phosphate reductoisomerase from *Aquilaria sinensis* (Lour.) Gilg. Journal of Genetics. 2015b;94(2):239–49.

Matsushita Y, Kang W, Charlwood BV. Cloning and analysis of a cDNA encoding farnesyl diphosphate synthase from *Artemisia annua*. Gene. 1996;172(2):207–9.

Oldfield S, Lusty C, MacKinven A. The world list of threatened trees. Cambridge, UK: World Conservation Press; 1998.

Paddon CJ, Keasling JD. Semi-synthetic artemisinin: a model for the use of synthetic biology in pharmaceutical development. Nat Rev Microbiol. 2014;12(5):355–67.

Persoon GA, Heuveling van Beek H. Growing 'the wood of the gods': agarwood production in southeast Asia. In: Snelder DJ, Lasco LD, editors. Smallholder tree growing for rural development and environmental services: lessons from Asia. Netherlands: Springer; vol. 5; 2008. p. 245–62.

Pojanagaroon S, Kaewrak C. Mechanical methods to stimulate aloes wood formation in *Aquilaria crassna* Pierre ex H.Lec. (Kritsana) trees. Acta Hortic. 2005;676:161–6. doi:10.17660/ActaHortic.2005.676.20.

Rasool S, Mohamed R. Plant cytochrome P450s: nomenclature and involvement in natural product biosynthesis. Protoplasma. 2015;1–13. doi:10.1007/s00709-015-0884-4.

Struck AW, Thompson ML, Wong LS, Micklefield J. S-adenosyl-methionine-dependent methyltransferases: highly versatile enzymes in biocatalysis, biosynthesis and other biotechnological applications. Chembiochem. 2012;13(18):2642–55.

Tholl D. Terpene synthases and the regulation, diversity and biological roles of terpene metabolism. Curr Opin Plant Biol. 2006;9(3):297–304.

Xu Y, Zhang Z, Wang M, Wei J, Chen H, Gao Z, Sui C, Luo H, Zhang X, Yang Y, Meng H, Li W. Identification of genes related to agarwood formation: transcriptome analysis of healthy and wounded tissues of *Aquilaria sinensis*. BMC Genomics. 2013a;14(1):227.

Xu Y, Yang X, Wei J, Zhang X, Zhang Z, Gao Z, Sui C. Cloning and expression analysis on 3-hydroxy-3-methylglutaryl-coenzyme A reductase from *Aquilaria sinensis*. Chinese Herbal Medicines. 2013b;3(5):182–8.

Xu YH, Yang X, Zhang Z, Liang L, Wei JH. Cloning and expression analysis of HMG-CoA reductase from *Aquilaria sinensis* (Lour.) Gilg. Yao Xue Xue Bao. 2013c;48(6):953–9.

Xu Y, Liu J, Liang L, Yang X, Zhang Z, Gao Z, Sui C, Wei J. Molecular cloning and characterization of three cDNAs encoding 1-deoxy-d-xylulose-5-phosphate synthase in *Aquilaria sinensis* (Lour.) Gilg. Plant Physiol Biochem. 2014;82:133–41.

Yagura T, Ito M, Kiuchi F, Honda G, Shimada Y. Four new 2-(2-phenylethyl)chromone derivatives from withered wood of *Aquilaria sinensis*. Chem Pharm Bull (Tokyo). 2003;51(5):560–4.

Yagura T, Shibayama N, Ito M, Kiuchi F, Honda G. Three novel diepoxy tetrahydrochromones from agarwood artificially produced by intentional wounding. Tetrahedron Lett. 2005;46(25):4395–8.

Yang X, Wei JH, Liu J, Xu YH. Cloning and expression analysis of farnesyl pyrophosphate synthase from *Aquilaria sinensis*. Zhongguo Zhong Yao Za Zhi. 2013;38(19):3251–5.

Zhang Z, Wei J, Han X, Liang L, Yang Y, Meng H, Xu Y, Gao Z. The sesquiterpene biosynthesis and vessel-occlusion formation in stems of *Aquilaria sinensis* (Lour.) Gilg trees induced by wounding treatments without variation of microbial communities. Int J Mol Sci. 2014a; 15(12):23589.

Zhang Z, Zhang X, Yang Y, Wei JH, Meng H, Gao ZH, Xu YH. Hydrogen peroxide induces vessel occlusions and stimulates sesquiterpenes accumulation in stems of *Aquilaria sinensis*. Plant Growth Regul. 2014b;72(1):81–7.

Chapter 6
Gyrinops walla: The Recently Discovered Agarwood-Producing Species in Sri Lanka

S.M.C.U.P. Subasinghe and D.S. Hettiarachchi

Abstract This chapter describes the morphology, distribution, wood anatomy, and variations of agarwood resin contents and resin content of *Gyrinops walla* endemic to Sri Lanka. We revealed for the first time, this species, which populates the lower elevations of the wet zone of Sri Lanka in 2012. More importantly, the recently identified species possesses agarwood-producing ability, similar to other species in the Thymelaeaceae family. Before this scientific discovery, *G. walla* was considered a least valuable species due to the very low stem density. Not much is known about this forgotten species; we intend to unleash its full potential as a new economic commodity to this country.

6.1 Introduction

Gyrinops walla is a tree endemic to Sri Lanka. Due to the minimal timber value and very low stem density, the villagers used to remove this tree from their homesteads; therefore it could only be observed in forests and along live fences. However, in 2012, this species became one of the most commonly discussed topics due to its ability to produce agarwood and the high selling prices associated with it published in the media. During that time, many smuggling efforts were caught by the authorities. Parallel to these activities, we conducted research on *G. walla*'s resin-producing ability (Subasinghe et al. 2012; Subasinghe and Hettiarachchi 2013). As far as our knowledge goes, these were inaugural reports on the species ability to form agarwood. Earlier scientific publications on *G. walla* describe its geographical distribution (Hou 1960) and chemical constituents in the leaves (Schun et al. 1986), which are different from that of agarwood.

S.M.C.U.P. Subasinghe (✉)
Department of Forestry and Environmental Science, University of Sri Jayewardenepura, Nugegoda, CO 10250, Sri Lanka
e-mail: upuls@sjp.ac.lk

D.S. Hettiarachchi
Wescorp Agarwood, Wescorp Group of Companies, 26, Coulson Way, Canning Vale, WA 6155, Australia

R. Mohamed (ed.), *Agarwood*, Tropical Forestry,
DOI 10.1007/978-981-10-0833-7_6

Furthermore, we proved the resins produced in *G. walla* are chemically similar to that of the commercially agarwood-producing species growing in Southeast Asia. Chemical analysis revealed the presence of several sesquiterpene compounds like agarofuran, vetispirane/agarospirol, prezizane, guaiane, eremophilane, and eudesmane/selinane (Subasinghe and Hettiarachchi 2013). We also conducted a study on seed germination, which saw a rapid decline in germination ability after the seeds fell to the ground. In addition, we are currently conducting agarwood resin formation studies using different fungal species following conventional treatment methods such as those adopted by Southeast Asian countries on *Aquilaria* species. In this chapter, we review the present scientific knowledge on *G. walla* including its distribution in Sri Lanka, wood anatomy, and chemical constituents and discuss issues related to its conservation and future prospects in Sri Lanka.

6.2 Botanical Description and Distribution

Gyrinops walla grows as a small to medium size tree (Gunatilleke et al. 2014) generally up to 15 m with a straight, slender but erect stem and a small rounded crown (Fig. 6.1a) (Dassanayake and Fosberg 1981). However, the crown shape can change from elongated to umbel depending on the position of the tree in the forest canopy and its growing stage. This tree bears numerous amounts of branches with slender and wiry twigs. The bark is smooth, thin, and strongly fibrous and its color varies from gray or brownish gray to reddish brown. *Gyrinops walla* leaves are alternatively arranged, slightly shinning and simple, and the buds are silky. Leaf shape is oblong and acute at the base. The average mature leaf size is 3.5×10 cm with a short rather abrupt bluntish acumen up to 1 cm long. Lateral veins of this species are very fine and numerous in number. The midrib is prominent. The petiole is short and 1–6 mm in length.

This tree bears bisexual flowers in the inflorescence. The flowers are small, slender, pubescent, and yellowish. Each flower bears 3–5 pedicels in shortly stalked umbels of axils. They contain tubular and slender perianth, which contains a ring of short hairs and scales above stamens. The ovary is superior with a pendulous ovule in each loculus. The fruit of *G. walla* is about 1.8 cm long capsule and obvate in shape and reddish brown, which bears two tadpole-like seeds in two valves. These seeds (Fig. 6.1b) are acuminate at the tip and covered with brown hairs and pointed (Dassanayake and Fosberg 1981; Jayaweera 1982).

Gyrinops walla is distributed in damper areas in the lower elevations of Sri Lanka, mainly in the humid lowland forests and home gardens of the southwest region (Fig. 6.1c). The elevation of this region is lower than 1000 m and the annual rainfall is between 2000 and 3000 mm. The average temperature is 25–27 °C. *Gyrinops walla* is also found in the moist forests of the central province of Sri Lanka where the elevation is higher than 1000 m, and the annual rainfall is lower than that of the southwest part of the country. However, we found that it is possible to observe this species in other climates (Subasinghe 2015).

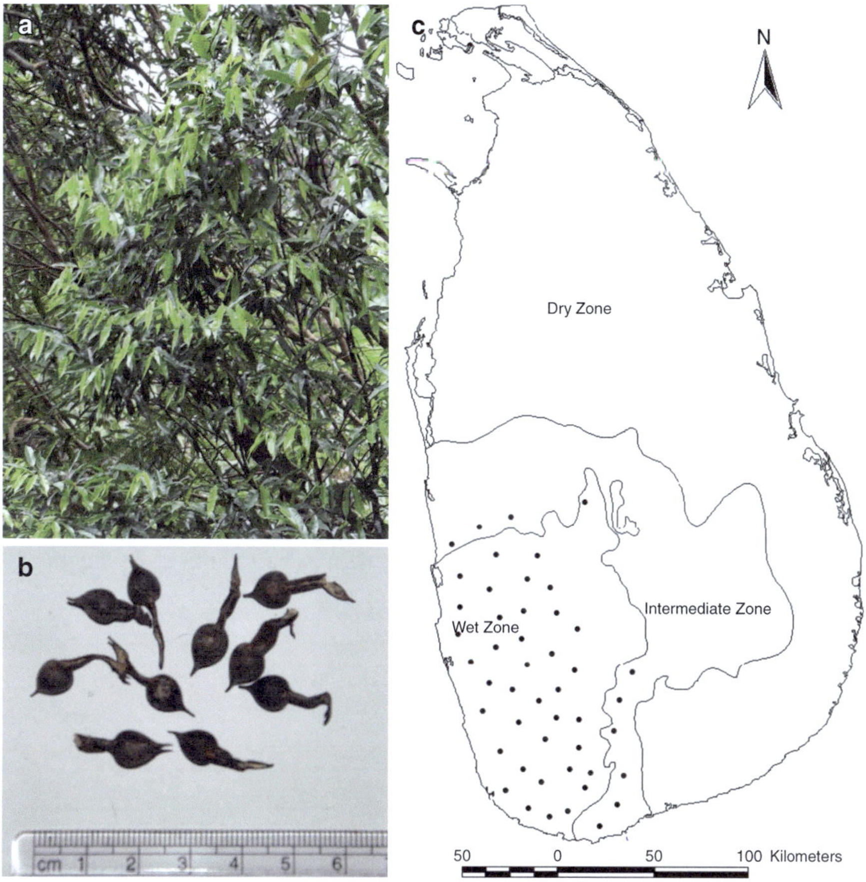

Fig. 6.1 Botanical features and distribution of *Gyrinops walla*. (**a**) Tree canopy, (**b**) seeds, and (**c**) distribution map in Sri Lanka

6.3 Traditional Use of the Tree

The fibrous bark of *G. walla* is well known for its durability. It can be easily stripped from the stem as narrow and long bands. Due to the strength of the bark, local people had used the bark strips to bind things like traditional medicinal pastes on broken limbs of human and large mammals such as buffaloes to help the healing process. Villagers also used the tender leaves of *G. walla* as vermifuges, especially for young children in those days when Western medicines were not popular for such purposes. Leaves were applied as poultice on boils and fistula, and in snake bites, often leading to recovery. Macerated leaves were also placed in the tooth cavities to loosen the infected teeth for easy removal. The tree has also been used for preparation of medicinal oils (Jayaweera 1982). The wood of *G. walla* is white or yellowish and very soft in nature. The stem wood density is 345 kg m^{-3} (Welivita and

Subasinghe 2006); therefore it has little value. The villagers used its wood for making handles for their gardening tools, but it has never been utilized for construction, furniture manufacturing, or similar uses in Sri Lanka.

6.4 Wood Anatomy

In our attempt to determine the species identity, we captured important elements of its wood anatomy under the microscope. Microscopic images of transverse, radial, and tangential sections of the tree stem are shown in Fig. 6.2a–c (unpublished data of the State Timber Corporation, Sri Lanka), respectively. *Gyrinops walla* has undefined growth rings in the stem (Fig. 6.2a). These growth rings are not clearly visible in the stem cross sections under the naked eye or even under the microscope. The main reason could be that this species is growing under favorable conditions throughout the year. The hallmark of trees from the Thymelaeaceae family including *Gyrinops* are the included phloems (Mohamed et al. 2013; Jiao et al. 2014), which are visible in the transverse section (Fig. 6.2a). The xylem vessels are mostly

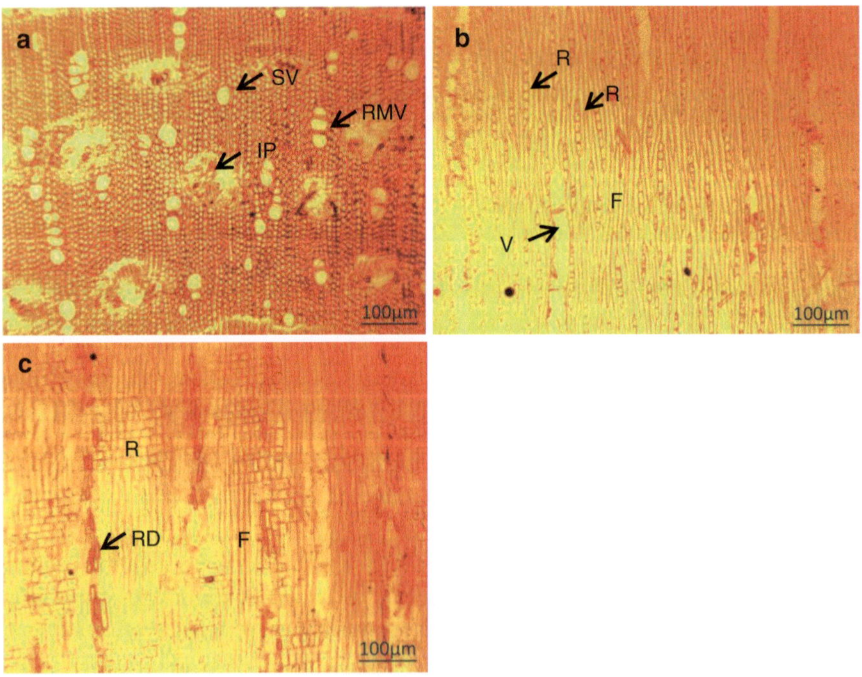

Fig. 6.2 Anatomical features of *Gyrinops walla*. (**a**) Transverse, (**b**) radial, and (**c**) tangential sections. Scale bars, 100 μm. *IP* included phloem, *SV* solitary vessel, *RMV* radial multiple vessel, *R* ray parenchyma, *F* fiber, *V* vessel, *RD* resin deposit (Source: State Timber Corporation of Sri Lanka)

radial and multiple, but a few solitary vessels are also present. The vacant areas, which are larger than the vessels, could be observed as the resin canals. The ray parenchymas are clearly visible in the tangential (Fig. 6.2b) and radial (Fig. 6.2c) sections. These living cells are possibly responsible in agarwood formation.

6.5 Chemical Constituents

In order to identify its quality, the agarwood produced in *G. walla* was heated on a hot piece of charcoal and a sensory panel assessed the odor. The top note was recorded as of sweet fruity aroma, followed by a short bodily middle note, and a lasting oriental woody base note. These odor notes were similar to what have been reported on agarwood of *Aquilaria* origins (Naef 2011); thus the use of *G. walla* as an agarwood source is confirmed.

Chemical constituents of *G. walla* have been reported by Schun and Cordell (1985) for its leaf constituents and their potential anticancer activity. Of the numerous compounds they extracted, two showed considerable activity against cancer cells. Furthermore, a triterpenoid named wallenone has been isolated and characterized from the leaves (Schun et al. 1986). The leaf, bark, and uninfected stem tissues were also studied using thin layer and gas chromatographic analysis (Dharmadasa et al. 2013). However, the current authors (Subasinghe et al. 2012; Subasinghe and Hettiarachchi 2013) were the first to characterize agarwood from *G. walla* and its chemical constituents. The initial attempt was to identify fragrant molecules previously reported in other agarwood-producing species such as from *Aquilaria* origin: *A. malaccensis* (synonym *A. agallocha*), *A. crassna*, and *A. sinensis*. Essential oil from agarwood contains sesquiterpene compounds and 2-(2-phenylethyl)-chromone derivatives and in some cases includes fatty acids (Naef 2011; Chen et al. 2012).

During the past 4 years, we studied over 150 individual *G. walla* trees containing agarwood. These trees were growing in various climatic regions of Sri Lanka. Chemical analysis of the above samples revealed that resins produced in *G. walla* have some major compound classes. However, the distribution and quantity varied between samples. This is not surprising because it is now known that agarwood formation depends upon several internal and external factors that affect the tree, including the type and time of attack. Those factors, however, do not affect the availability of major compound classes (Xu et al. 2013). Chen et al. (2012) have reported six different classes of sesquiterpenes from agarwood, while Naef (2011) has made a more elaborative classification of eight classes.

Our studies have identified six classes of sesquiterpenes from *G. walla* (Table 6.1). Agarofuran, agarospirol, and jinkohol types have been reported by several groups (Maheshwari et al. 1963; Varma et al. 1965; Nakanishi et al. 1981), while eudesmanes and guaiane types have been reported in a series of studies conducted by Ishihara et al. (1991a, b, 1993a, b); all were described from *Aquilaria* species. Apart from the sesquiterpenes, 4-phenyl-2-butanone has been identified as a major

Table 6.1 Sesquiterpene molecules identified in *Gyrinops walla* via GC-MS

Sesquiterpene class	Compound
Agarofuran type	Agarofuran
Vetispirane type/agarospirane type	Agarospirol
Vetispirane type/agarospirane type	Baimuxinic acid
Vetispirane type/agarospirane type	*oxo*-Agarospirol
Vetispirane type/agarospirane type	*Unknown compound*[a]
Vetispirane type/agarospirane type	*Unknown compound*[a]
Vetispirane type/agarospirane type	*Unknown compound*[a]
Prezizane type	Jinkohol
Prezizane type	Jinkoh-eremol
Guaiane type	Azulenone
Guaiane type	*Unknown compound*[a]
Guaiane type	Aromadendrene
Guaiane type	Isolongifolene
Guaiane type	Alloaromadendrene oxide
Guaiane type	Guaia-(10),11-dien-15-ol
Guaiane type	*Unknown compound*[a]
Eremophilane type	9,1-Eremophiladien-8-one
Eremophilane type	Isopropyl naphthalene (derivative)[a]
Eremophilane type	Isopropyl naphthalene (derivative)[a]
Eudesmane/selinane type	γ-Eudesmol
Eudesmane/selinane type	Valerenol
Eudesmane/selinane type	γ-Elemene
Eudesmane/selinane type	2,2,6,8-Tetramethyl bicyclo undece-7-en-3-ol
Eudesmane/selinane type	β-Selinene
Eudesmane/selinane type	Eudesmane-4-(14), 11-diene
Eudesmane/selinane type	Selina-3,11-diene-9-one
Eudesmane/selinane type	Selina-3,11-diene-14-al

[a]Mass spectroscopic and Kovat's indices matched to the sesquiterpene class but could not be configured to a particular compound

aromatic compound, which is present in every agarwood sample analyzed during our studies; these aromatic ketones have been reported from agarwood of *Aquilaria* origin (Chen et al. 2012). Upon analysis of the dichloromethane extract, we identified two 2-(2-phenylethyl) – chromones. Further investigations into chromone compounds are necessary as they have been identified as potential markers for identifying agarwood origins (Espinoza et al. 2014) and thus could be applied to *G. walla*.

In our research, the retention indices (RIs) of *G. walla* constituents were compared to that of authentic *A. crassna* (Wetwitayaklung et al. 2009), *A. sinensis* (Chen et al. 2012), and *A. agallocha* (Nor Azah et al. 2008) (Table 6.2) (Subasinghe and Hettiarachchi 2013). The results confirmed the similarity of agarwood resins produced in *G. walla* with that of the tested *Aquilaria* species.

Table 6.2 Retention indices (RIs) of several compounds identified in *Gyrinops walla* compared to *Aquilaria* species using a 5 % phenyl and 95 % methyl siloxane capillary column (DB-5 type)

Compound	*G. walla*	*A. crassna*	*A. sinensis*	*A. agallocha*
Jinkoh-eremol	1641	1643	–	1650
Selina-3,11-diene-9-one	1689	1687	–	–
Selina-3,11-diene-14-al	1733	1735	1733	–
9,11-Eremophiladien-8-one	1741	1740	–	–
Guaia-(10),11-dien-15-ol	1766	1770	–	–
oxo-Agarospirol	1818	1822	–	–

Source: Subasinghe and Hettiarachchi (2013)

6.6 Conservation Status

6.6.1 Illegal Harvesting

Due to the facts that products from naturally formed agarwood enjoy high prices in the world market and supply of such agarwood from Southeast Asian countries is decreasing, agarwood from the virgin *G. walla* resources in Sri Lanka has drawn a great deal of attention from high-end users and traders. To make matters worse, the lucrative financial figures published in the media also attracted illegal harvesters, which significantly enhanced the extent of poaching. Illegal harvesting activity reached its highest peak within the last 2 years. Poachers cut the trees without having prior knowledge on the availability of agarwood within the stems. This act had sacrificed a large number of trees in the rainforests of the country. Once the tree populations of such forests declined, poachers turned to homestead-grown trees. According to published media information, within 2013, the police and customs of Sri Lanka had confiscated 13,800 kg of processed agarwood chips. However, the real figure could be much higher because essential oils extracted from the chips have not been seized.

6.6.2 Present Legal Status

The numbers of *G. walla* quickly dwindled in Sri Lanka after 2012. Apparently, the high monetary value of *G. walla* made publicly by both electronic and printed media had attracted illegal harvesters who poached agarwood. In addition, villagers also cut *G. walla* trees in search of agarwood resin inside the stems. Because of lack of knowledge about agarwood formation and its subsistence, people cut trees of all sizes causing a severe threat on the existence of this species. Consequently, the authorities captured many smugglers, both local and foreign. In December 2012, the Biodiversity Secretariat of Sri Lanka recategorized *G. walla* as "vulnerable" species, after which the export of timber, tissues, or any extracts from the species was banned.

6.7 Germination and Propagation

There was very little research focusing on *G. walla*. Furthermore, cultivation and plantation were not established in the past because the species had not been recognized as having any commercial value. Propagation of this species was made in the traditional way, and no attempts were made on seed germination until we conducted the first trial under funding from the National Research Council of Sri Lanka in 2013. We found that the germination rate increased up to 55 % when seeds were sown in pure coarse sand with good draining. To maximize germination success, the seeds must be germinated within the first 2 weeks after dropping from the trees because their viability declines rapidly. The use of different seed treatments may enhance the germination rate. One difficulty faced by researchers is the low availability of *G. walla* trees due to the heavy poaching. In addition, the small-sized green fruits are well camouflaged in the tree canopies, resulting in difficult fruit harvesting. The ripened fruits attract birds such as the hornbills, which feed on them.

6.8 Agarwood Production Ability of *G. walla*

We tested agarwood-producing ability of *G. walla* in trees growing under natural conditions and in three different populations, i.e., Ingiriya, Mirigama, and Yagirala, all in the low country wet zone of Sri Lanka. The wet zone seems to favor *G. walla*'s growth and the abundance was high at the time our research started. To increase the distance between the three populations, the areas for sample collection were selected from three different administrative districts within the Western Province. Diameters of the selected trees varied from 8 to 21 cm and the heights varied from 10 to 24.5 m.

Careful observations were made in identifying stems or branches that contain naturally formed agarwood. Once located, such resinous wood was manually separated from the healthy wood, and the particle size was manually reduced before extracting with dichloromethane using the cold maceration method. Chemical analysis was conducted using gas chromatography–mass spectrometry (GC-MS) as commonly applied in essential oil analysis. Extracts were dissolved in anhydrous acetone before injected into the instrument. Narrow-bore capillary columns were used with both 5 % phenyl siloxane and polyethylene glycol inner coating attached to a guard column. From our preliminary testings, we found that the 5 % phenyl siloxane column was more efficient at separating different classes of compounds. Separated compounds were identified by comparing the fragmentation patterns of mass spectroscopy and Kovat's indices of published data (Nor Azah et al. 2008; Wetwitayaklung et al. 2009; Chen et al. 2012).

The resin contents varied from 0.68 to 3.8 % for the three populations, with averages between 1.8 and 2.4 % (Fig. 6.3); however the values were insignificantly different. The results proved for the first time that *G. walla* produces agarwood

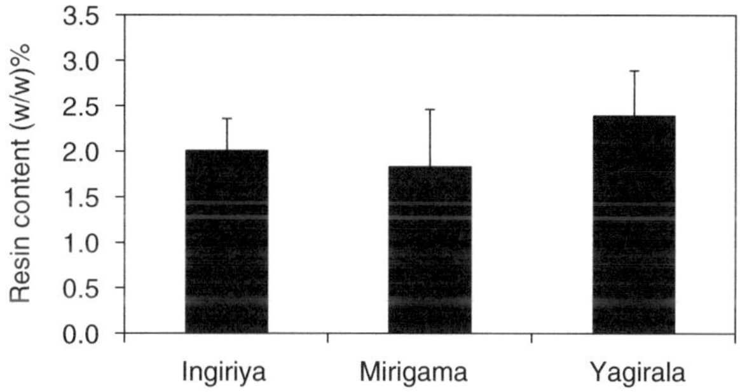

Fig. 6.3 Mean in resin content within the selected *Gyrinops walla* populations (±SE)

resins due to natural reasons. However, destructive sampling was not conducted during this experiment, and therefore the agarwood volume as a percentage to the entire tree volume was not estimated.

The three populations have shown a variation in the distribution of resin constituents. Most of the samples analyzed contained key agarwood aroma compounds such as agarofuran, agarospirol, phenyl butanone, jinkohol, guaiane-type sesquiterpenes, and chromones (Table 6.3). Mirigama sample was rich in sesquiterpenes but phenyl butanone was absent, whereas agarofuran was absent from Ingiriya sample. The most prominent sesquiterpenes were jinkohol, alloaromadendrene oxide, and azulenone, whereas some vetispirane-type sesquiterpenes were found only in a few samples. These variations could occur because of geographic variations among the selected populations, resin age, and other factors that might have played significant roles in resin variations.

During our studies, we have developed a gas chromatographic fingerprint for *G. walla* resin (Fig. 6.4 and Table 6.3), which could be compared between *G. walla* samples and against agarwood of other species. Chemical fingerprint of *G. walla* was compared to that of several authentic agarwood samples obtained from *A. malaccensis* and *A. crassna*. The distribution of major sesquiterpenes responsible for aroma could also be compared based on the fingerprint where peak retention was confirmed by Kovat's indices. In accordance with the results, agarwood produced in *G. walla* has similar quality to commercial agarwood available in the market. Future chemical analysis would elucidate the absolute chemical identities of key compounds in *G. walla*, which could be compared with reports from other species. These marker compounds will be used as a tool for quality assurance of *G. walla* as a more sustainable source of agarwood.

6.9 Current and Future Research

There is very limited information on agarwood production in *G. walla* when compared to its closely related species from *Aquilaria* origin. This is expected because of the recently discovered status of this species in Sri Lanka. Realizing the importance of *G. walla* to the economy of this country, currently we are conducting studies targeting at commercializing *G. walla* through plantation establishment for

Table 6.3 Main compounds and compound types identified in *Gyrinops walla* via GC-MS

No	Compound	Type	Ingiriya	Mirigama	Yagirala
1	4-Phenyl-2-butanone	NA	2.52 ± 0.85	0.00	1.83 ± 0.86
2	Agarofuran	Agarofuran	0.00	3.22 ± 1.21	0.46 ± 0.19
3	Agarospirol	Vetispirane	1.10 ± 0.65	0.59 ± 0.29	0.90 ± 0.33
4	Unknown	Vetispirane	0.00	0.00	1.88 ± 0.90
5	*oxo*-Agarospirols	Vetispirane	1.90 ± 0.72	0.51 ± 0.21	2.40 ± 0.86
6	Unknown	Vetispirane	0.71 ± 0.41	0.00	0.88 ± 0.35
7	Baimuxinic acid	Vetispirane	0.00	1.46 ± 1.03	1.05 ± 0.34
8	Jinkohol	Prezizane	6.5 ± 0.28	13.71 ± 3.60	5.36 ± 1.34
9	Azulenone	Guaiane	3.48 ± 0.60	9.61 ± 3.24	8.48 ± 2.87
10	Aromadendrene	Guaiane	0.00	1.20 ± 0.69	1.06 ± 0.39
11	Isolongifolene	Guaiane	1.00 ± 0.58	3.33 ± 0.53	2.38 ± 0.58
12	Alloaromadendrene oxide	Guaiane	2.96 ± 0.16	14.80 ± 3.99	7.31 ± 1.03
13	9,1-Eremophiladien-8-one	Eremophilane	0.00	3.49 ± 2.04	3.01 ± 1.44
14	Guaia-(10),11-dien-15-ol	Guaiane	0.59 ± 0.34	0.00	1.44 ± 0.55
15	Eudesmane-4-(14), 11-diene	Eudesmane/selinane	1.09 ± 0.69	0.73 ± 0.33	2.45 ± 0.94
16	β-Selinene	Eudesmane/selinane	0.67 ± 0.39	4.8 ± 2.43	0.51 ± 0.22
17	Octadecanoic acid	Fatty acids	1.13 ± 0.72	7.69 ± 2.92	3.63 ± 1.62
18	2-(2-Phenylethyl)-chromone derivative	2-(2-Phenylethyl)-chromone	1.00 ± 0.88	10.88 ± 7.49	3.63 ± 1.83
19	2-(2-Phenylethyl)-chromone derivative	2-(2-Phenylethyl)-chromone	14.99 ± 2.20	4.55 ± 4.24	7.26 ± 2.26

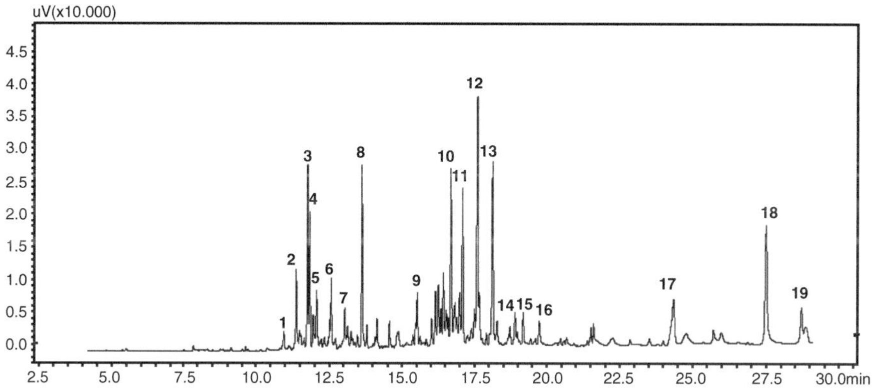

Fig. 6.4 A representative gas chromatographic fingerprint of *Gyrinops walla* extract. *Number* on each peak matches the compound listed in Table 6.3

agarwood cultivation. This research is jointly funded by the government of Sri Lanka and the private sector. The objective is to identify optimal practices in seed germination and nursery handling for improving seed germination rates and survival of *G. walla* seedlings because seedlings are in high demand for starting a plantation.

To cultivate agarwood, different conventional methods are being tested on *G. walla*. In addition, fungal species colonizing naturally formed agarwood tissues and soils where *G. walla* communities are present have already been isolated and identified. They include members of *Trichoderma*, *Fusarium*, *Aspergillus*, and *Diplodadium* from the agarwood tissues, and *Mucor*, *Aspergillus*, *Fusarium*, *Dendrophoma*, *Trichoderma*, and *Absidia* from the forest soils. It is expected that some of these fungal inhabitants of agarwood tissues and forest soils are capable to enhance agarwood formation due to natural reasons. To test their ability, several isolated species have been inoculated into *G. walla*, and the physical and chemical properties of the trees are being measured at present.

Differences in resin content and constituents will be compared between artificially and naturally formed agarwood using standard hydro-distillation method and GC-MS. Our previous testings found that the ionization method used in GC-MS was intense and resulted with similar fragmentation patterns to related compounds. Therefore, absolute identification of derivatives of a certain class of compounds was not possible. In the future, a tandem liquid chromatography with mass spectrometry will be experimented. This method will provide the ability to analyze nonvolatile compounds such as 2-(2-phenylethyl)-chromone derivatives as reported by Lancaster and Espinoza (2012). Other methods such as GC-FTIR and NMR spectroscopic analysis will also be tested in the future. In addition, we are currently developing an optimized thin-layer chromatography (TLC) method for *G. walla*, which has thus far not been reported. The aim of a TLC method is to quantitatively and qualitatively asses the resin with reference to major chemical compound classes. Due to the high monetary value, agarwood resin extraction methods have to be optimized. Three critical parameters, i.e., soaking time, pressure, and distillation durations, will be tested for this purpose.

6.10 Conclusion and Future Perspectives

Our studies proved that the quality of agarwood produced in *G. walla* was similar to that produced by *Aquilaria* species from Southeast Asian regions. *Gyrinops walla* contains all key compounds found in *Aquilaria* species, including 2-(2-phenylethyl) chromones which are unique to agarwood (compounds 18 and 19 in Table 6.3). This places *G. walla* at par with other known agarwood-producing species in the world.

The key activities in agarwood trade are nursery and plantation establishment, agarwood cultivation, resin extraction, and product manufacturing. Unfortunately, *G. walla* plantations are not available in Sri Lanka. Due to the present policy of the government that does not permit oil extraction from *G. walla* for export, and complicated transport issues, plantation establishment is not popular among the villagers or private sectors. In addition, agarwood oil extraction factories or relevant industries have not been established in the country. The present unclear legal status has become a constraint to implement income generation opportunities. Agarwood plantation is considered a short-term business owing to the brief rotation cycle when compared with other forestry-related investments such as teak and mahogany plantations. Because agarwood is a fast income-generating business, the private sectors in Sri Lanka have started introducing *Aquilaria* plantations as short-term forestry projects, which are gaining popularity since the last 2 years. In the future, more large- and medium-scale *Aquilaria* plantations will be established in Sri Lanka rather than the endemic *G. walla* if the legal barriers are not relaxed. It is also not clear if export will be allowed for *G. walla* trees grown in private lands even under the strict supervision of the government officials such as the Forest Department or Department of Wildlife Conservation. Ironically, the government has branded agarwood from *G. walla* as "Sri Lankan agar." Although this is a positive step in creating a new image and foreseeably demands for Sri Lankan agarwood in the international market, proper regulations have to follow, which promote cultivation of agarwood from this species as well as protecting its natural population in the forest. If export of agarwood from *G. walla* is permitted, government intervention is still necessary in creating proper market channels. Extensive research is needed in product manufacturing as to diversify the agarwood industry in Sri Lanka, from supplying raw materials to manufacturing and exporting the end products to rich nations. This would be a more sustainable approach to boost the economy of the people and the country.

References

Chen HQ, Wei JH, Yang JS, Zhang Z, Yang Y, Gao ZH, Gong B. Chemical constituents of agarwood originating from the endemic genus *Aquilaria* plants. Chem Biodiv. 2012;9:236–50.

Dassanayake MD, Fosberg FR. Flora of Sri Lanka, vol. 2. New Delhi: Oxford and IBH Publishing Company; 1981.

Dharmadasa R, Siriwardana A, Samarasinghe K, Adhihetty P. Standardization of *Gyrinops walla* Gaertn. (Thymalaeaceae): newly discovered, fragrant industrial potential, endemic plant from Sri Lanka. World J Agric Res. 2013;1:101–3.

Espinoza EO, Lancaster CA, Kreitals NM, Hata M, Cody RB, Blanchette RA. Distinguishing wild from cultivated agarwood (*Aquilaria* spp.) using direct analysis in real time and time of-flight mass spectrometry. Rapid Commun Mass Spectrom. 2014;28:281–9.

Gunatilleke IAUN, Punchi-Manage R, Fernando RHSS, Gunatilleke CVS. Spatial distribution and characteristics of *Gyrinops walla* Gaertn (Thymalaeaceae) In Sri Lanka. Proceedings of 18th international forestry and environment symposium, University of Sri Jayewardenepura, Sri Lanka; 2014.

Hou D. Thymelaeaceae. In: Van Steenis CGGJ, editor. Flora Malesiana series I, vol. 6. Netherlands: Wolters-Noordhoff Publishing; 1960. p. 1–48.

Ishihara M, Tsuneya T. Components of the volatile concentrate of agarwood. J Essen Oil Res. 1993;5:283–9.

Ishihara M, Tsuneya T, Shiga M, Uneyama K. Three sesquiterpenes from agarwood. Phytochemistry. 1991a;30:563–6.

Ishihara M, Tsuneya T, Uneyama K. Guaiane sesquiterpenes from agarwood. Phytochemistry. 1991b;30:3343–7.

Ishihara M, Tsuneya T, Uneyama K. Fragrant sesquiterpenes from agarwood. Phytochemistry. 1993;33:1147–55.

Jayaweera DMA. Medicinal plants (indigenous and exotic) used in Ceylon: part V. The National Foundation of Sri Lanka; 1982.

Jiao L, Yin Y, Cheng Y, Jiang X. DNA barcoding for identification of the endangered species *Aquilaria sinensis*: comparison of data from heated or aged wood samples. Holzforschung. 2014;68(4):487–94.

Lancaster C, Espinoza E. Evaluating agarwood products for 2(2-phenylethyl) chromones using direct analysis in real time time-of-flight mass spectrometry. Rapid Commun Mass Spectrom. 2012;26:2649–56.

Maheshwari ML, Jain TC, Bates RB, Bhattacharyya SC. Structure and absolute configuration of alpha-agarofuran, beta-agarofuran and dihydroagarofuran. Tetrahedron. 1963;19:1079–90.

Mohamed R, Wong MT, Halis R. Microscopic observation of ‘gaharu’ wood from *Aquilaria malaccensis*. Pertanika J Technol Sci. 2013;36(1):43–50.

Naef R. The volatile and semi-volatile constituents of agarwood, the infected heartwood of *Aquilaria* species: a review. Flavour Fragrance J. 2011;26:73–87.

Nakanishi T, Yamagata E, Yoneda K, Miura I. Jinkohol, a prezizane sesquiterpene alcohol from agarwood. Phytochemistry. 1981;20:1597–9.

Nor Azah MA, Chang YS, Mailina J, Abu Said A, Abd Amjid JSHS, Nor Hasinida H, Nik Yasmin Y. Comparison of chemical profiles of selected gaharu oils from Peninsular Malaysia. Malays J Anal Sci. 2008;12:338–40.

Schun Y, Cordell GA. Studies in the Thymelaeaceae III. Constituents of *Gyrinops walla*. J Nat Prod. 1985;48:684–5.

Schun Y, Cordell GA, Cox PJ, Howie RA. Wallenone, a C_{32} triterpenoid from the leaves of *Gyrinops walla*. Phytochemistry. 1986;25:753–5.

Subasinghe SMCUP. Aromatic oils: The neglected resource to strengthen the national economy. Vidya. 2015;17(2):3–4.

Subasinghe SMCUP, Hettiarachchi D. Agarwood resin production and resin quality of *Gyrinops walla* Gaertn. Int J Agric Sci. 2013;3:357–62.

Subasinghe SMCUP, Hettiarachchi DS, Rathnamalala E. Agarwood-type resin from *Gyrinops walla* Gaertn: a new discovery. J Trop Forest Environ. 2012;2(2):43–8.

Varma KR, Maheshwari ML, Bahattacharyya SC. The constituents of agarospirol, a sesquiterpenoid with a new skeleton. Tetrahedron. 1965;21:115–38.

Welivita I, Subasinghe SMCUP. Estimation of above ground biomass of trees at Yagirala Forest Reserve. Proceedings of the 26th Annual Sessions of the Institute of Biology, Sri Lanka; 2006.

Wetwitayaklung P, Thavanapong N, Charoenteeraboon J. Chemical constituents and antimicrobial activity of essential oil and extracts of heartwood of *Aquilaria crassna* obtained from water distillation and supercritical fluid carbon dioxide extraction. Silpakorn University Sci Technol J. 2009;3:25–33.

Xu Y, Zhang Z, Wang M, Wei J, Chen H, Gao Z, Sui C, Luo H, Zhang X, Yang Y. Identification of genes related to agarwood formation: transcriptome analysis of healthy and wounded tissues of *Aquilaria sinensis*. BMC Genomics. 2013;14:227.

Chapter 7
Resolution of Complex Sesquiterpene Hydrocarbons in *Aquilaria malaccensis* Volatile Oils Using Gas Chromatography Technique

Saiful Nizam Tajuddin, Che Mohd Aizal, and Mashitah Mohd Yusoff

Abstract Agarwood or *gaharu* is resin-impregnated wood of the tree genus *Aquilaria* (Thymelaeaceae). In Malaysia, the main agarwood producer is *Aquilaria malaccensis* and oil extracted from this species is highly priced. One of the challenges in commercializing agarwood is the lack of universal standard to classify the aromatic oils. Our present knowledge places the main aromatic compounds of agarwood oil in the sesquiterpene hydrocarbon region. In this work, we extracted agarwood oil using hydrodistillation method in the laboratory and compared with a commercial-scale extraction in the factory. We analyzed the sesquiterpene hydrocarbon region using several highly sophisticated detection systems. Using GC-FID, 12 sesquiterpene hydrocarbons were identified, while another eight were determined using GC-MS. Five compounds were identified in both analytical techniques: aromadendrene, α-bulnesene, α-guaiene, γ-gurjunene, and β-maaliene. Advanced analysis using GC×GC/TOFMS detected 24 sesquiterpene hydrocarbons in both laboratory and pilot scale agarwood oils. Many of the sesquiterpene hydrocarbons identified provide the woody aroma to the agarwood oil. Specifically, α-gurjunene and α-guaiene contribute to the woody balsamic aroma, while α-copaene contributes to the spicy-wood aroma. In total, 33 sesquiterpene hydrocarbons were identified from *A. malaccensis* in the present study, with high certainty. Results from this study can be used toward establishing a universal standard for agarwood oil from the genus *Aquilaria* in the global market, which is presently lacking.

S.N. Tajuddin (✉) • C.M. Aizal
Bio Aromatic Research Center of Excellence, Universiti Malaysia Pahang, Gambang, Pahang 26300, Malaysia
e-mail: saifulnizam@ump.edu.my

M.M. Yusoff
Faculty of Industrial Sciences and Technology, Universiti Malaysia Pahang, Gambang, Pahang 26300, Malaysia

R. Mohamed (ed.), *Agarwood*, Tropical Forestry,
DOI 10.1007/978-981-10-0833-7_7

7.1 Introduction

Natural agarwood is the pathological product from diseased *Aquilaria* tree primarily because of wounds on the trunk. Consequently, the tree produces a type of resin (known as agar, agarwood, or *gaharu*) that is both high in volatile organic constituents and fragrant as a response to the attack. It is believed that the fragrant resin assists the tree in suppressing or retarding the microbial growth. The affected wood became dark brown or black due to the increased mass and wood density from resin development, leaving the unaffected wood in its original pale beige color. The high-quality agarwood is normally recognized from its dark color and strong aroma. The term black agar refers to the agarwood that resembles black stone as indicated in Fig. 7.1. Agarwood is primarily used as incense, while its essential oil is in heavy demand in the perfume industry as evidenced from the recent expansion of new consumer products such as agarwood essence, soap, and shampoo.

Several methods have been used to classify the physical properties of agarwood for trade purposes. For instance, brown agarwood without any black tone is often used as incense. Meanwhile, yellow wood with interspersed bands of brown or black resin is graded as the lowest quality and is commonly distilled for the oil (Chang et al. 2002). Therefore, low-cost agarwood oil can be obtained to be utilized in toiletries or perfumeries or blended as carrier oils. At present, there is very little information on the quality of different agarwood essential oils produced.

Essential oils, otherwise known as essences or volatile oils, are the volatile of secondary metabolites produced by plants either for their nutrition or other purposes (e.g. as protectant or attractant). The "essential oil" term is often used to illustrate the complex mixtures of chemical components in plants generally extracted using steam distillation, solvent extraction, or physically pressed plant material techniques (Ernest 1948; Yeung 1980). Their chemical constituents are widely utilized for their fragrance, flavor, pharmacological, antimicrobial, insect repellency, and other medicinal properties (Bakkali et al. 2008; Burfield and Reekie 2005; Burt 2004; Edris 2007).

Fig. 7.1 A high-quality natural agarwood often has a dark and dense resinous appearance

Minor differences in these oil compositions can significantly alter their odor or flavor. Therefore, to ensure consistency of the products manufactured over time, the extraction process and composition analysis must be done precisely. Often time, the identification of numerous components in essential oils has been very challenging due to their complex mixture (Adams 2001). Nowadays, several analytical techniques have been developed to solve separation and identification of complex mixture compounds in essential oils for instant enantioseparation. Therefore, this chapter intends to explore the resolution mixture of sesquiterpene hydrocarbons in high-quality agarwood oil using gas chromatography (GC) and advanced two-dimensional GC system.

7.2 Hydrodistillation for Agarwood Essential Oil Extraction

There are several methods of essential oil extraction that have been applied in research laboratories and industries such as the hydrodistillation, microwave-assisted, and supercritical fluid extraction methods. Each method has its own advantages and disadvantages in relation to the quality of the extracted oil such as the yield, duration, cost, and ability to extract targeted compounds (Augusto et al. 2003; Boris 2005; Luque de Castro et al. 1999).

However, hydrodistillation is the most popular method for extracting agarwood essential oil in commercial scale due to easy operation, lower cost, and green water-based process. The copper distillation pot has been traditionally used and it is still in use by some traders. Nowadays, the copper has been replaced by stainless steel. In this method, the pot is filled with fermented agarwood chips or powder and then heated on a brick furnace (Fig. 7.2a). The heating process takes a long time, possibly 2–3 continuous days to ensure all the oil has been extracted. The hot vapor then goes through a steel condenser, where external running water jacket cools the vapors,

Fig. 7.2 A commercial hydrodistillation system for agarwood oil extraction in Gua Musang, Kelantan (Malaysia). (**a**) Extraction contraption and (**b**) extracted oil in the collecting cone

which then drop into a collecting separating funnel where oil separates from water gravitationally (Fig. 7.2b) (Burfield and Kirkham 2005).

Meanwhile, laboratory scale hydrodistillation technique utilizes a Clevenger-type apparatus in which the raw material is heated to reflux as illustrated in Fig. 7.3. The essential oil components tend to form an azeotropic mixture with water during this process, and vapors containing volatile constituents are carried along with evaporated steam to a condenser. Prior to the extraction, distilled water is introduced into Clevenger apparatus with 5 ml of analytical grade hexane or other nonpolar solvents via the inlet. At the end, the mixture of essential oil and hexane is collected and mixed with anhydrous sodium sulfate to remove water content. Essential oil then is purified by passing through inert gas usually nitrogen.

The laboratory hydrodistillation period can take from 3 to 24 h depending on the sample. Distillation temperatures should be 100 °C; however, the biomass movement in the pot and heat distribution can vary and produce temporary high temperature, which in turn can lead to formation of artifacts. A well-regulated operation system can help in reducing compound rearrangement and terpene decomposition. Oil of poor quality can also be produced as the result of prolonged heating in contact with water (Stewart 2005). The polymerization of aldehydes, hydrolysis of esters, or decomposition (e.g., dehydration) of volatile components commonly occurs during this stage.

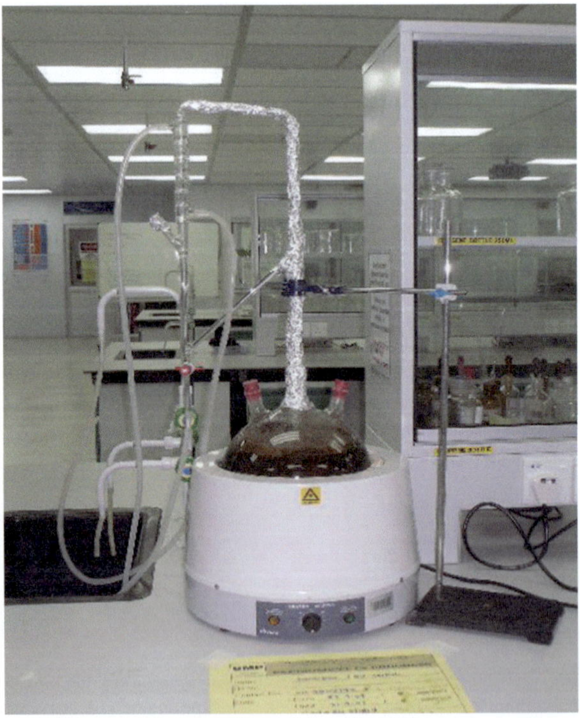

Fig. 7.3 Hydrodistillation using Clevenger-type apparatus

7.3 Components of Volatile Agarwood Oil

Agarwood contains several types of sesquiterpenes, and it has become the subject of active research in the past 40 years or more. It is believed that Agarol was the first sesquiterpene isolated from agarwood (Bhattacharyya et al. 1952; Maheshwari et al. 1963; Bucchi and Wuest 1979). Nowadays, more than 70 sesquiterpene compounds have been identified from agarwood (Naef 2011). The basic molecule of terpenes is the isoprene unit C_5H_8. Chemically, the terpenes can be divided into several classes including mono-, sesqui-, and diterpenes. Monoterpenes are comprised of two isoprene units, $C_{10}H_{16}$, whereas sesquiterpenes, $C_{15}H_{24}$, contained three isoprene units. Sesquiterpenes that are important in agarwood oils are sesquiterpene hydrocarbons ($C_{15}H_{24}$) and oxygenated sesquiterpenes ($C_{15}H_{26}O$). They are derived from the sesquiterpene skeletons and grouped as agarofuran, agarospirane/vetispirane, cadinane, eremophilane/valencane, eudesmane/selinane, guaiene, and prezizane types (Fig. 7.4). Important oxygenated sesquiterpenes including agarospirol, jinkohol, jinkohol-eromol, and kusenol that may contribute to the characteristic woody aroma of agarwood are shown in Fig. 7.5.

Generally, agarwood oils are mixtures of sesquiterpenes, oxygenated sesquiterpenes, oxygenated phenyls, carboxylic, and carbonyl hydrocarbons. Table 7.1 shows the main sesquiterpene hydrocarbons and oxygenated sesquiterpenes, which have been identified by previous agarwood research.

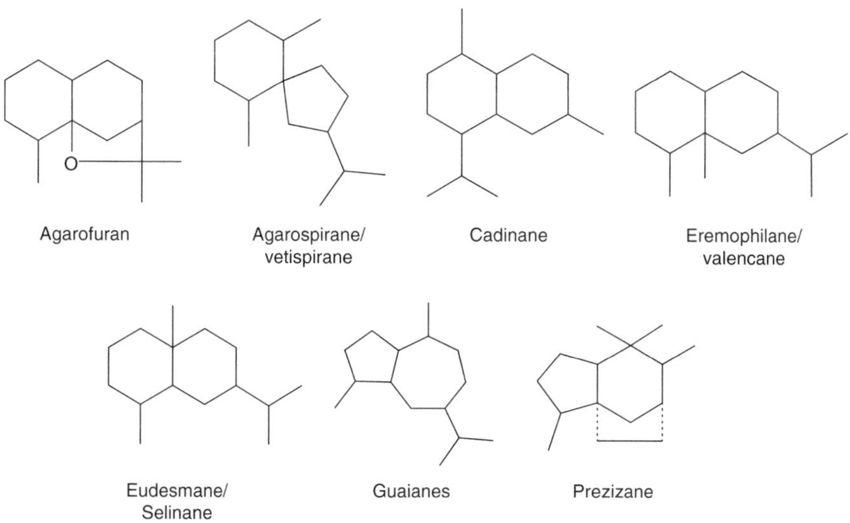

Fig. 7.4 Skeleton of hydrocarbons found in volatile agarwood oil (Source: Naef (2011), Chen et al. (2012))

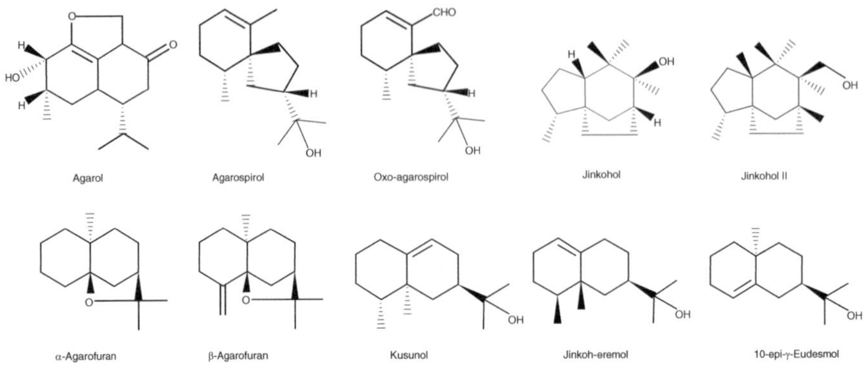

Fig. 7.5 Skeleton of oxygenated sesquiterpenes found in volatile agarwood oil (Source: Maheshwari et al. (1963), Varma et al. (1965), Nakanishi et al. (1981, 1984), Ishihara and Tsuneya (1993a, b))

7.4 Compound Detection via Gas Chromatography (GC)

GC is a well-known analytical technique utilized for separation of volatile compounds. It provides qualitative and quantitative information for each compound present in the sample. Initially, the solid or liquid sample is vaporized into a gaseous state by heating process. As the vaporized sample moves through a GC column, these compounds are fractioned between a mobile phase (gas) and a stationary phase (solid or liquid). The compounds are separated in time and space due to the differential fractionation of the solute into the stationary phase.

GC finds its main application in the analyses of essential oils, fatty acids, mono- and sesqui-terpenes, and sulfur compounds (Arian 2002; Bartle and Myers 2002). An inert carrier gas under pressure (helium, argon, nitrogen, or hydrogen) is used to convey the vaporized sample through a narrow column. Normally, GC columns are made from fused silica tubes with internal diameters between 0.1 and 1 mm. To enhance thermal stability, 0.1–5 μm stationary phase films are bound and cross-linked to the column's inner surface. The column then is stationed inside an oven with temperature control. This allows the column to be heated slowly from an ambient temperature up to 350–450 °C for separating the chemical compounds.

Flame ionization detector (FID) and mass spectrometer detector (MSD) are the most commonly used instruments for detection of eluting compounds in GC. The applicability of GC with "universal" detectors makes this analytical instrument one of the most suitable option for essential oil study. GC has three main advantages over other analytical methods; it is very rapid, it has a very high separation capacity, and it is highly sensitive (König et al. 1999).

Table 7.1 List of sesquiterpenes from agarwood oils of four *Aquilaria* species from different countries and extraction mode

Aquilaria species	Extraction mode	Main constituents	Reference
A. agallocha from:			
Bangladesh (Sylhet)	Hydrodistillation	Aristolene; caryophyllene oxide; γ-eudesmol; β-selinene; valencene	Bhuiyan et al. (2009)
India	Petroleum ether	α-Agarofuran; β-agarofuran; *nor*-ketoagarofuran; 3,4-dihydroxydihydro-agarofuran	Maheshwari et al. (1963)
Indonesia (type B)	Benzene	Jinkohol II	Nakanishi et al. (1983)
Indonesia (type B)	Benzene	Agarol; oxo-agarospirol; (-)-10-*epi*-eudesmol	Nakanishi et al. (1984)
Vietnam (Kyara, Kanankoh)	Acetone	Selina-3,11-dien-9-one; selina-3,11-dien-14-al	Ishihara et al. (1991a)
Vietnam (Ryoku-yu, Kanakoh)	Acetone	(+)-Guaia-1(10),11-dien-15, 2-olide; (-)-guaia-1(10),11-dien-15-ol; rotundone	Ishihara et al. (1991b)
Vietnam (type A)	Benzene	β-Agarofuran; *nor*-ketoagarofuran; agarospirol; oxo-agarospirol; dihydrokaranone; jinkoh-eremol; kusunol	Yoneda et al. (1984)
A. crassna from:			
Thailand	Hydrodistillation, supercritical CO_2	Agarospirol; γ-eudesmol; 10-*epi*-γ-eudesmol; epoxy bulnesene; kusunol; selina-3,11-dien-9-one; selina-3,11-dien-14-al; γ-selinene;	Wetwitayaklung et al. (2009)
Thailand (cultivated, Bo Rai in Trat province)	Hydrodistillation (pretreatment with enzyme), supercritical CO_2	Agarospirol; caryophyllene oxide; eudesma-4,11-diene-3-one; 10-*epi*-γ-eudesmol; guaia-3,9-diene; β-guaiene; selina-3,7(11)-diene; δ-selinene	Yoswathana (2013)
Thailand (cultivated, Bo Rai in Trat province)	Hydrodistillation with different pretreatment, subcritical water extraction	Agarospirol; aristol-9-dien-8-one; eudesma-4,11-dien-3-one; guaia-3,9-diene; α-guaiene; β-guaiene; nootkatone; selina-3,7(11)-diene	Yoswathana et al. (2012)
Thailand (Trat province)	Hydrodistillation	α-Agarofuran; agarospirol; aromadendrene epoxide; aristolene; α-gurjunene; 10-*epi*-γ-eudesmol	Pornpunyapat et al. (2011)

(continued)

Table 7.1 (continued)

Aquilaria species	Extraction mode	Main constituents	Reference
A. malaccensis from:			
India (Hojai in Assam province)	Hydrodistillation	Aromadendrene; (+)-calarene; 6-guaiadiene; valencene	Jayachandran et al. (2014)
Indonesia (type B)	Benzene	Jinkohol	Nakanishi et al. (1981)
Indonesia (type B)	Benzene	Jinkohol II	Nakanishi et al. (1983)
Indonesia (type B)	Benzene	Agarol, 3,4-dihydroxydihydro-agarofuran, oxo-agarospirol, jinkoh-eremol	Nakanishi et al. (1984)
Malaysia (grade C)	Hydrodistillation	α-Agarofuran; β-agarofuran; *nor*-ketoagarofuran; agarospirol; 3-phenyl-2-butanone; β-eudesmol; 10-*epi*-γ-eudesmol; α-guaiene; jinkoh-eremol; jinkohol II; kusunol	Nor Azah et al. (2008)
Vietnam (Kyara, Kanankoh)	Acetone	Oxo-agarospirol, α-guaiene, (-)-Guaia-1(10),11-dien-15-al	Ishihara et al. (1991a)
Vietnam (type A)	Benzene	α-Agarofuran, 3,4-dihydroxydihydro-agarofuran, oxo-agarospirol, jinkoh-eremol, jinkohol, jinkohol II, kusunol	Yoneda et al. (1984)
A. sinensis from			
China	Hydrodistillation	Agarospirol; caryophyllene oxide; α-copaen-11-ol; eremophila-7(11),9-diene-8-one; eudesm-7(11)-en-4α-ol; γ-eudesmol; guaia-1(10),11-dien-9-one; guaiol; selina-3,11-dien-14-al; α-selinene	Chen et al. (2011)
China (Hainan, Guangdong)	Ethanol, hydrodistillation	Agarospirol; aristolene; caryophyllene oxide; α-copaen-11-ol; eudesm-7(11)-en-4α-ol; γ-eudesmol; guaiol	Xing et al. (2012)

Source: Tajuddin (2010), Naef (2011)

7.4.1 Gas Chromatography-Flame Ionization Detector (GC-FID)

GC-FID is a common technique to identify compounds by comparison of retention time (t_r). Those retention times are converted into Kovats retention index (I), which is a system-independent constant. The I values are actually the normalized of t_r value for adjacently eluting n-alkanes with those t_r values obtained from chromatogram. Parameters such as film thickness and diameter, column length, void time, carrier gas velocity, and pressure influenced the t_r values. However, derivation of I values is relatively independent, thus allowing the comparison of values calculated under various conditions and different analytical laboratories. Joulain and König (1998) suggested that by comparing experimental I values with those known values, Kovats indices can be used to identify chemical compounds in the samples.

The sensitive detection of flame ionization detector (FID) toward molecules with carbon-hydrogen bonds makes it the most utilized detector in GC. The low limit of detection ranging from 0.1 ppm to almost 100 % is highly preferable for hydrocarbon analysis. FID responds poorly to compounds such as CCl_4, H_2S, and NH_3 and in some cases not at all. Nonetheless, changing the carrier gas flow rate only has small impact on detector response. As a matter of fact, FID is not concentration sensitive, but it is sensitive to mass of hydrocarbons. The stable response rate also resulted in FID being more resistant to contamination either by column bleed or sample itself. Although it is a user-friendly and robust instrument, the sample is destroyed during analysis as compound ionization process uses hydrogen diffusion flame.

7.4.2 Gas Chromatography-Mass Spectrometry (GC-MS)

GC-MS is literally known as a hyphenated analytical technique as it combines two techniques, namely, GC and mass spectrometry (MS) to form a powerful single method for chemical compound analysis including essential oils. This combination is very useful for qualitative and quantitative evaluation of known and unknown sample as GC works to separate the component of a solution and MS characterizes each compound individually.

Mass spectrometry detection is based on analysis of generated charged particles (ions) from molecules of the analytes. This technique provides details regarding their molecular weight and chemical structures. Generally, various MS detector types and sample introduction techniques enable a broad application for sample analysis. According to Gross (2004), there are three specific regions for MS instrumentation; ionizer, ion analyzer, and detector.

7.4.3 Comprehensive Two-Dimensional Gas Chromatography (GC × GC)

GC × GC technique nowadays is utilized for its capability to analyze very complex samples (Marriott et al. 2001; Beens and Brinkman 2005; Adahchour et al. 2006; Mondello et al. 2007). This analytical tool is so powerful that it is extensively used to deal with separation and resolution issues in conventional GC. Giddings (1984) was the first person known to discuss and elucidate many basic requirements of "separation dimension" in his revolutionary paper on chromatography techniques. A single column for separation based on particular parameters (stationary phase, length, size, etc.) represents a separation dimension. Therefore, the term "GC × GC" indicates that two different columns are employed to create a two-dimensional separation for the sample. In fact, separation process in these columns is orthogonal and independent of each other. Despite the fact that the concept of GC × GC separation was first suggested by Giddings as a possibility, the initial practical implementation was demonstrated in 1991 (Marriott and Shellie 2002).

The extensive implementation and rapid improvement of GC × GC primarily contributed by its significant enhancement in peak capacity, sensitivity, resolution, as well as selectivity of the separation, aside for well-ordered and highly structured chromatograms. Originally, GC × GC was mainly utilized for petroleum products analysis. However, forensic science and food and fragrance industry have also reported to successfully use of GC × GC for respective studies (Dimandja et al. 2000; Micyus et al. 2005; Mondello et al. 2005, 2007; Khummueng et al. 2006; Min et al. 2006; Ma et al. 2007; Rocha et al. 2007; Cardeal and Marriott 2009; Pripdeevech et al. 2010). According to Schoenmakers et al. (2000), the formerly unachievable results using one dimension (1DGC), regardless if it was combined with the most powerful MS detectors, can be obtained through GC × GC analysis.

7.5 A Case Study of Essential Oil Analysis of Agarwood from Peninsular Malaysia

We conducted this study using the same source of agarwood samples, namely, *A. malaccensis*. Both infected woods used in the laboratory and commercial-scale hydrodistillation originated from the Kelantan Forest.

7.5.1 Essential Oil Extraction

Parameters used in the extraction procedures and characteristics of the oil produced are listed in Table 7.2. Prior to the extraction, wood samples were grinded into sawdust to maximize surface area and thus enhancing oil yield.

Table 7.2 Comparisons between oils extracted using laboratory and commercial-scale methods

	Laboratory	Commercial
Raw material, kg	0.1	25
Duration of extraction, h	12	36
Yield, %	0.2	0.2
Optical rotation, $[\alpha]^{25}$	−0.1065	−0.1084
Odor characteristic	Sweet, woody, aromatic	Sweet, woody, aromatic
Color	Greenish	Dark green

7.5.2 Compound Analysis Using GC-FID and GC-MS

GC-FID analysis was performed using an Agilent 7890A Network System gas chromatography, while GC-MS with the same system was attached to a mass spectrometer (Agilent 5975C) with detector in full-scan mode under electron impact ionization (EI, 70 eV). A capillary column (DB-1 ms 30 m×0.25 mm I.D.; 0.25 μm film thickness) purchased from J&W Scientific (Folsom, California) was chosen for the compound analysis by GC-FID and GC-MS. The oven temperature was programmed for 60 °C for 1 min and then ramped at 3 °C/min to 250 °C and held for 10 min. Injector inlet and detector temperatures were set at 250 °C, while sample injection (1 μL) was set at split ratio 1:5.

Over 36 components have been identified from the oil samples (Table 7.3). The laboratory oil had 36.7 % carboxylic acid derivatives, 18 % sesquiterpene hydrocarbons, and 31.9 % oxygenated sesquiterpenes, while the commercial oil had 3.4 %, 9.4 %, and 34.6 %, respectively. We identified eight sesquiterpene hydrocarbons in both oil samples (Table 7.3). Major sesquiterpene hydrocarbons in the laboratory oil were α-guaiene (5.8 %), β-selinene (4.9 %), and α-muurolene (3.4 %) and in the commercial oil α-guaiene (2.8 %) and α-bulnesene (2.8 %) (Figs. 7.6 and 7.7).

7.5.3 Compound Analysis Using Comprehensive GC×GC System

An Agilent 6890 GC, a Pegasus IV time-of-flight mass spectrometer (LECO Corp., St. Joseph, MI, USA), and a cold-jet modular KT-2001 Retrofit prototype (Zoex Corp., Lincoln, NE, USA) were used for GC×GC analysis. The first column was a nonpolar DB-1 ms (30 m×0.25 mm I.D.; 0.25 μm film thickness) and the second column was a DB-Wax (1.0 m×0.10 mm I.D.; 0.10 μm film thickness). Both columns were purchased from J&W Scientific (Folsom, California). Helium was used as carrier gas with head pressure at 37 atm. The initial temperature of the first dimension was maintained at 60 °C for 1 min, and the subsequent temperature program was ramped at a rate of 3 °C min^{-1} until 220 °C where it was held isothermally for 10 min. Meanwhile, the initial temperature of the second dimension was 75 °C,

Table 7.3 Chemical composition of volatile agarwood oils based on gas chromatography analysis

Compounds	DB1 column	Laboratory-extracted oil[a]	Commercial hydrodistilled oil[a]	Identification[b, c]
Carboxylic acid derivatives				
Benzaldehyde	935	3.3	–	RI, MS
2-Hydroxy-benzaldehyde	1003	0.6	–	RI, MS
Acetophenone	1066	0.7	–	RI, MS
4-Phenyl-2-butanone	1210	32.1	3.4	RI, MS
Total, %		**36.7**	**3.4**	
Sesquiterpene hydrocarbons				
β-Maaliene	1414	0.4	0.7	RI, MS
α-Guaiene	1440	5.8	2.8	RI, MS
Aromadendrene	1443	–	0.6	RI, MS
γ-Gurjunene	1472	0.7	1	RI, MS
β-Selinene	1486	4.9	–	RI, MS
α-Muurolene	1496	3.4	0.7	RI, MS
γ-Guaiene	1499	1.5	0.8	RI, MS
α-Bulnesene	1503	1.3	2.8	RI, MS
Total, %		**18**	**9.4**	
Oxygenated sesquiterpenes				
α-Elemol	1530	–	3.3	RI, MS
Caryophyllene oxide	1600	0.9	8.6	RI
Guaiol	1603	0.6	1.2	RI
Humulene epoxide II	1606	1.7	2.3	RI
1,5-Epoxy-*nor*-ketoguaiene	1614	1.1	0.6	RI, MS
10-*epi*-γ-Eudesmol	1619	1.6	0.8	RI, MS
Agarospirol	1631	0.9	1.4	RI, MS
epi-α-Cadinol	1640	–	2.9	RI
Jinkoh-eremol	1643	6.5	0.5	RI
Kusunol	1650	1	0.6	RI
α-Eudesmol	1652	0.7	0.9	RI, MS
Bulnesol	1664	1.5	0.6	RI, MS
Dehydrojinkoh-eremol	1673	1.4	1.2	RI, MS
epi-α-Bisabolol	1678	1.5	1	RI
α-Bisabolol	1683	1.8	0.5	RI
Selina-3,11-dien-9-one	1687	1.3	0.5	RI, MS
Rotundone	1703	0.5	–	RI, MS
Guaia-1(10),11-dien-15-al	1806	1.7	–	RI, MS
Guaia-1(10),11-dien-15-oic acid	1811	0.6	–	RI, MS
Karanone	1812	–	1.1	RI, MS
Oxo-agarospirol	1822	0.3	0.8	RI, MS

Table 7.3 (continued)

Compounds	DB1 column	Laboratory-extracted oil[a]	Commercial hydrodistilled oil[a]	Identification[b, c]
Eudesmol	1880	2.1	3.2	RI, MS
n-Hexadecanoic acid	1948	4.2	2.1	RI
Guaia-1(10),11-dien-15,2-olide	2019	–	0.5	RI, MS
Total, %		**31.9**	**34.6**	

Source: Tajuddin (2010)

[a]Components are listed in order of their relative content >0.1 %

[b]RI, linear retention indices were determined relative to the retention times on a DB-1 column of a homologous series of C_8-C_{20} *n*-alkanes

[c]MS, identification by comparison of the MS with those of the NIST library (>90 % match from the library)

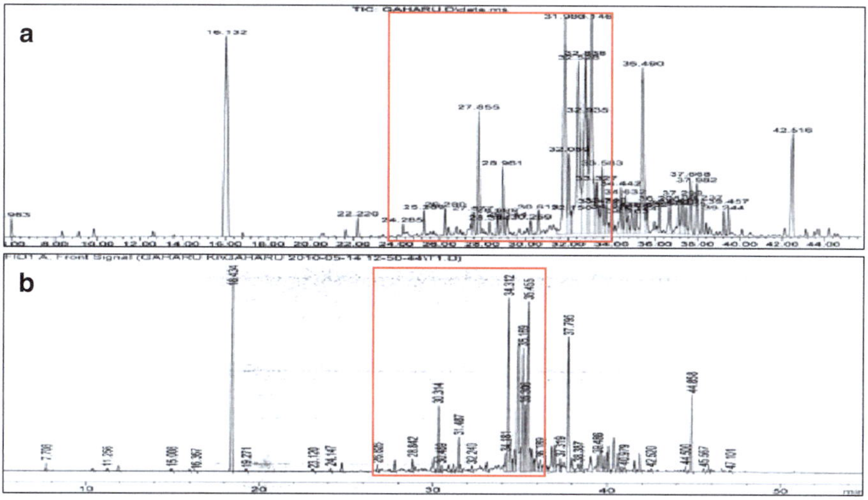

Fig. 7.6 Sesquiterpene hydrocarbon positions in (**a**) GC-FID and (**b**) GC-MS chromatograms, for laboratory oil

held for 30s, and then heated at 5 °C min^{-1} until 230 °C and again held isothermally for 10 min. Peak identification was made using TOFMS with electron impact ionization.

Eighteen sesquiterpene hydrocarbons were identified in the laboratory oil (Table 7.4, Fig. 7.8a) and 12 in the commercial oil (Table 7.5, Fig. 7.8b), bringing to a total of 24 in both oil samples (Table 7.6). Major sesquiterpene hydrocarbons in the laboratory sample were α-bulnesene (15.7 %), α-caryophyllene (13.61 %), and α-guaiene (11.31 %), whereas the commercial sample was dominated by aromadendrene (27.7 %), β-cubebene (9.5 %), and α-guaiene (8.7 %).

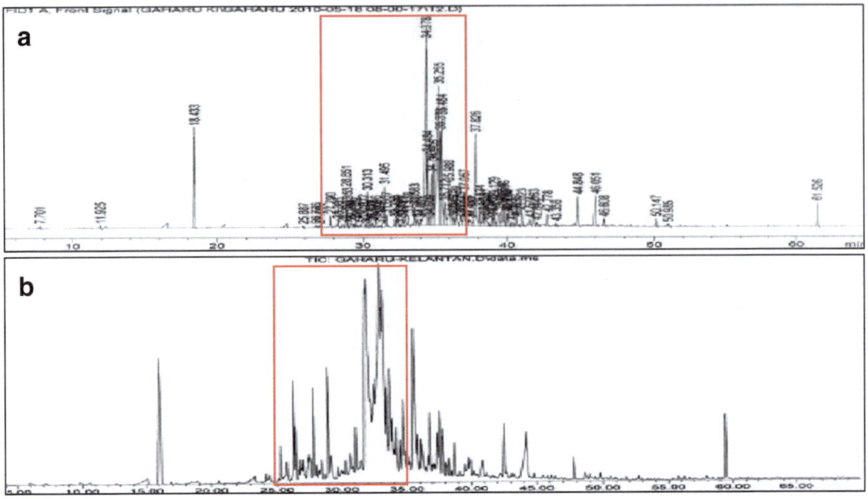

Fig. 7.7 Sesquiterpene hydrocarbon positions in (**a**) GC-FID and (**b**) GC-MS chromatograms, for commercial oil

7.5.4 General Discussion

Sesquiterpene hydrocarbons in agarwood oil are poorly studied although they are main contributors to the aroma arising from plant materials. For example, β-elemol is a fragrant sesquiterpene, which contributes to peppery and lemony odors. Other sesquiterpene like β-caryophyllene contributes to sweet, spicy, and fruity odors (Breitmaier 2006). A total of 27 sesquiterpene hydrocarbons were identified using GC-FID, GC-MS, and advance analysis via GC×GC/TOFMS (Table 7.7). GC-MS and GC-FID analyses found only seven to eight sesquiterpene hydrocarbons in the agarwood oil samples, whereas GC×GC/TOFMS was detected between 12 and 18. Most of these compounds also exhibit a specific odor unique to agarwood oil (Table 7.8).

Commercial oil contains lesser content of carboxylic acid derivatives in comparison to laboratory oil and is due to vaporization during the prolonged extraction time of 3 days (Tajuddin and Yusoff 2010). Previously, four compounds have been reported from agarwood, β-gurjunene (Ishihara and Tsuneya 1993a; Wetwitayaklung et al. 2009), γ-selinene (Wetwitayaklung et al. 2009), α-guaiene (Ishihara et al. 1991b), and α-bulnesene (Ishihara and Tsuneya 1993a; Wetwitayaklung et al. 2009), similar to our findings.

Due to extensive isomerization, sesquiterpene hydrocarbons tend to present themselves as a complex mixture in many types of essential oil analyses (Shellie et al. 2001). Conventional techniques such as GC-MS are not efficient in resolving such a complex mixture of sesquiterpene hydrocarbons (Fig. 7.9a). Improvements to the multidimensional gas chromatography (GC×GC) have enabled resolution to

Table 7.4 Sesquiterpene hydrocarbons identified in the laboratory oil via GC × GC/TOFMS

Peak	Compounds	R.T. (s)	Similarity	Reverse	Probability	Area %	CAS
1	Isoledene	990, 2.060	801	801	1824	0.22	95910-36-4
2	α-Copaene	980, 1.990	909	911	4590	0.35	3856-25-5
3	Selina-3,7(11)-diene	1335, 2.570	648	651	3712	0.21	6813-21-4
4	δ-Cadinene	1305, 2.460	850	856	3435	1.74	483-76-1
5	α-Guaiene	1265, 2.470	934	934	4680	11.31	3691-12-1
6	*allo*-Aromadendrene	1240, 2.480	885	887	1678	2.46	25246-27-9
7	α-Selinene	1225, 2.400	884	898	1939	3.82	473-13-2
8	α-Muurolene	1210, 2.360	895	901	4741	2.49	31983-22-9
9	α-Himachalene	1160, 2.330	803	831	1410	0.22	3853-83-6
10	α-Caryophyllene	1150, 2.360	930	934	7663	13.61	6753-98-6
11	Aromadendrene	1145, 2.330	847	857	1839	10.28	109119-91-7
12	β-Patchoulene	1130, 2.250	901	908	3358	5.64	514-51-2
13	α-Bulnesene	1115, 2.240	947	950	3124	15.69	3691-11-0
14	Germacrene D	1100, 2.230	820	865	1329	0.26	23986-74-5
15	β-Caryophyllene	1075, 2.260	929	929	3250	1.19	87-44-5
16	β-Sesquiphellandrene	1065, 2.230	883	898	1060	0.62	20307-83-9
17	*trans*-α-Bergamotene	1065, 2.110	870	875	4787	2.04	13474-59-4
18	di-*epi*-α-Cedrene	1035, 2.140	879	881	1140	0.14	50894-66-1

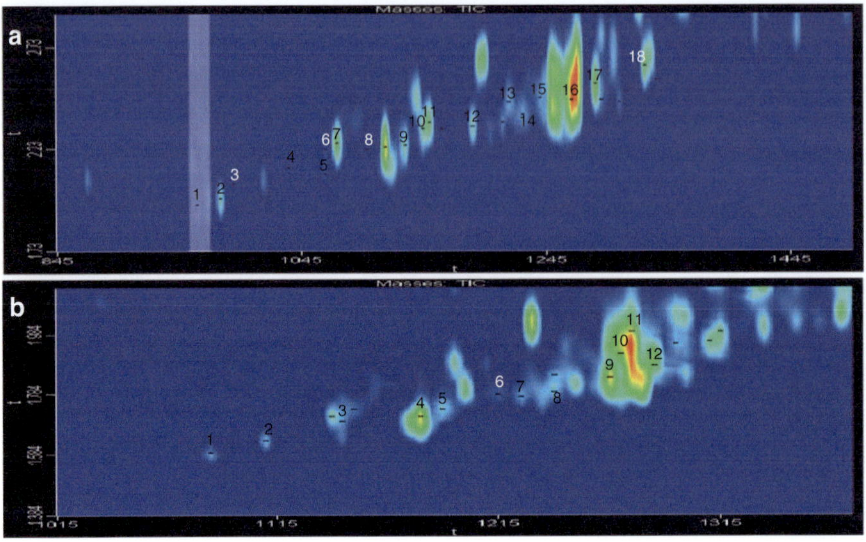

Fig. 7.8 Partial 2D GC×GC/TOFMS chromatogram for the (**a**) laboratory and (**b**) commercial oils. Labeled compounds correspond to the respective peaks listed in Tables 7.4 and 7.5

Table 7.5 Sesquiterpene hydrocarbons identified in the commercial oil via GC×GC/TOFMS

Peak	Compounds	R.T. (s)	Similarity	Reverse	Probability	Area %	CAS
1	α-Copaene	1085, 1.592	919	921	5087	0.94	3856-25-5
2	β-Elemene	1110, 1.632	900	900	6237	1.25	33880-83-0
3	α-Cedrene	1140, 1.712	911	914	5092	4.45	469-61-4
4	β-Caryophyllene	1150, 1.736	903	903	2881	0.87	87-44-5
5	α-Guaiene	1180, 1.712	948	948	4351	8.73	3691-12-1
6	β-Patchoulene	1190, 1.736	905	917	4462	1.88	514-51-2
7	α-Longipinene	1215, 1.784	832	847	2210	0.82	5989-08-2
8	*allo*-Aromadendrene	1225, 1.776	807	824	923	2.55	25246-27-9
9	α-Muurolene	1240, 1.792	893	908	4652	4.13	31983-22-9
10	α-Curcumene	1240, 1.848	914	927	9149	4.01	644-30-4
11	Aromadendrene	1270, 1.920	836	836	3395	27.66	109119-91-7
12	β-Cubebene	1285, 1.880	809	829	2904	9.45	13744-15-5

Table 7.6 Sesquiterpene hydrocarbons identified via GC × GC/TOFMS

No.	Compounds	Laboratory oil	Commercial oil	CAS registration
1.	*allo*-Aromadendrene	√	√	25246-27-9
2.	Aromadendrene	√	√	109119-91-7
3.	*trans*-α-Bergamotene	√	–	13474-59-4
4.	α-Bulnesene	√	–	3691-11-0
5.	δ-Cadinene	√	–	483-76-1
6.	α-Caryophyllene	√	–	6753-98-6
7.	β- Caryophyllene	√	√	87-44-5
8.	γ-Caryophyllene	–	–	118-65-0
9.	di-*epi*-α-Cedrene	√	–	50894-66-1
10.	α-Cedrene	–	√	469-61-4
11.	α-Copaene	√	√	3856-25-5
12.	β-Cubebene	–	√	13744-15-5
13.	α-Curcumene	–	√	644-30-4
14.	β-Elemene	–	√	33880-83-0
15.	Germacrene D	√	–	23986-74-5
16.	α-Guaiene	√	√	3691-12-1
17.	α-Himachalene	√	–	3853-83-6
18.	Isoledene	√	–	95910-36-4
19.	α-Longipinene	–	√	5989-08-2
20.	α-Muurolene	√	√	31983-22-9
21..	β-Patchoulene	√	√	514-51-2
22.	Selina-3,7(11)-diene	√	–	6813-21-4
23.	α-Selinene	√	–	473-13-2
24.	β-Sesquiphellandrene	√	–	20307-83-9

the complex mixture (Fig. 7.9b). In such cases, structural identification was attained at high certainty (Wu et al. 2004). The ability of GC×GC to enhance peak capacity when combined with deconvolution power from the TOFMS system greatly improves quantification (Fig. 7.9c).

Our results demonstrate that a higher potential in compound detection and identification can be achieved via GC×GC/TOFMS. Using this approach, the number of detected sesquiterpene hydrocarbons was doubled when compared to GC-FID and GC-MS. Better analyte separation via GC×GC and enhanced sensitivity (full mass range acquisition) of TOFMS detection contribute to superior identification capacity (Marriott et al. 2001; Mondello et al. 2007). GC×GC/TOFMS is also applicable for detection of co-eluted peaks due to its deconvolution algorithm and automated peak acquiring. In addition, duration and complexity of the analysis are reduced with the presence of two different polarity columns and established software library for GC×GC.

Table 7.7 Distribution of sesquiterpene hydrocarbons identified in volatile agarwood oils

No.	Compounds	Laboratory oil			Commercial oil		
		GCFID	GCMS	GC×GC/TOFMS	GCFID	GCMS	GC×GC/TOFMS
1.	*allo*-Aromadendrene	√	–	√	√	–	√
2.	Aromadendrene	–	√	√	–	√	√
3.	*trans*-α-Bergamotene	–	–	√	–	–	–
4.	α-Bulnesene	√	√	√	√	√	–
5.	δ-Cadinene	–	–	√	–	–	–
6.	α-Caryophyllene	–	–	√	√	–	–
7.	β- Caryophyllene	–	–	√	–	–	√
8.	γ-Caryophyllene	–	–	–	–	–	–
9.	di-*epi*-α-Cedrene	–	–	√	–	–	–
10.	α-Cedrene	–	–	–	–	–	√
11.	α-Copaene	–	–	√	–	–	√
12.	β-Cubebene	–	–	–	–	–	√
13.	α-Curcumene	–	–	–	–	–	√
14.	β-Elemene	–	–	–	–	–	√
15.	Germacrene D	–	–	√	–	–	–
16.	α-Guaiene	–	√	√	–	√	√
17.	γ-Gurjunene	–	√	–	√	–	–
18.	α-Himachalene	–	–	√	–	–	–
19.	Isoledene	–	–	√	–	–	–
20.	α-Longipinene	–	–	–	–	–	√
21.	β-Maaliene	√	√	–	–	√	–
22.	α-Muurolene	–	–	√	–	–	√
23.	β-Patchoulene	–	–	√	–	–	√
24.	Selina-3,7(11)-diene	–	–	√	–	–	–
25.	α-Selinene	–	–	√	–	–	–
26.	β-Selinene	–	√	–	–	–	–
27.	β-Sesquiphellandrene	–	–	√	–	–	–

Table 7.8 List of major compounds identified in agarwood oil and their odor characteristics

Compounds	Odor characteristics[a]
allo-Aromadendrene	Wood
Aromadendrene	Wood
trans-α-Bergamotene	Wood
α-Bulnesene	Wood
α-Cedrene	Wood
α-Caryophyllene	Wood
α-Copaene	Wood, spice
α-Guaiene	Wood, balsamic
β-Gurjunene	Wood, balsamic
α-Muurolene	Wood
α-Selinene	Wood

[a]Odor characteristics are based on Terry and Heinrich (1984)

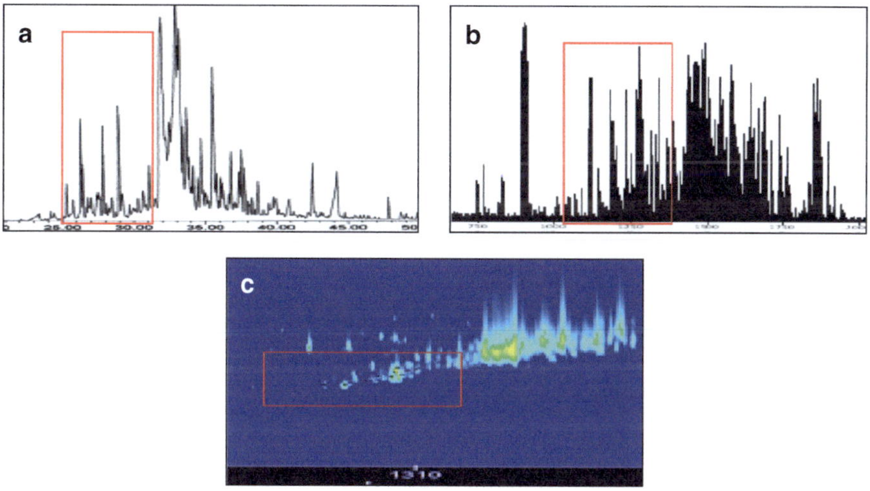

Fig. 7.9 GC × GC/TOFMS has the ability to resolve complex mixtures of sesquiterpene hydrocarbons in volatile agarwood oil. (**a**) GC-MS chromatogram (unable to resolve), (**b**) 1D GC × GC/TOFMS chromatogram (able to resolve), and (**c**) 2D chromatogram GC × GC/TOFMS

7.6 Conclusions

We found GC × GC/TOFMS an effective tool for separating chemical components in very complex mixtures such as in the volatile oils of agarwood, far greater than when using conventional GC-FID or GC-MS. When coupled to TOFMS, the GC × GC system shows a marked improvement in sensitivity and resolution. TOFMS has higher reverse values and similarity compared to MS detector, in addition to its superior mass spectral data in the reference library. This allows identification of a higher number of peaks in the agarwood oils. GC × GC/TOFMS could be adopted as a useful tool to meet demands in agarwood oil grading for trade and forensic purposes.

Acknowledgments The authors acknowledge the Ministry of Science, Technology and Innovation (MOSTI) of Malaysia for financial support under the e-Science Fund (02-01-16-SF0092). GC-FID, GC-MS, and GC × GC/TOFMS facilities were provided by Universiti Malaysia Pahang and Universiti Kebangsaan Malaysia. Special thanks to Mazlan Mohamed for knowledge sharing of the agarwood industry.

References

Adahchour M, Beens J, Vreuls RJJ, Brinkman UAT. Recent developments in comprehensive two-dimensional gas chromatography, further applications, conclusion and perspectives. Trends Anal Chem. 2006;25(8):821–39.

Adams RP. Identification of essential oil components by gas chromatography/quadrupole mass spectroscopy. Carol Stream: Allured Publishing Corp; 2001.

Arian VA. The importance of GC and GC-MS in perfume analysis. TrAC Trends Anal Chem. 2002;21:698–708.
Augusto F, Lopes AL, Zini CA. Sampling and sample preparation for analysis of aroma and fragrances. Trends Anal Chem. 2003;22(3):160–9.
Bakkali F, Averbeck S, Averbeck D, Idaomar M. Biological effects of essential oils – a review. Food Chem Toxicol. 2008;46(2):446–75.
Bartle KD, Myers P. History of gas chromatography. Trends Anal Chem. 2002;21:547–57.
Beens J, Brinkman UAT. Comprehensive two-dimensional gas chromatography – a powerful and versatile technique. Analyst. 2005;130(2):123–7.
Bhattacharyya B, Datta A, Baruah HK. On the formation and development of agar in *Aquilaria agallocha*. Sci Cult. 1952;18:240–1.
Bhuiyan MZI, Begum J, Bhuiyan MNH. Analysis of essential oil of eaglewood tree (*Aquilaria agallocha* Roxb.) by gas chromatography mass spectrometry. Bangladesh J Pharm. 2009;4:24–8.
Boris LM. Identification of chemical compounds. TrAC Trends Anal Chem. 2005;24:493–508.
Breitmaier E. Terpenes, flavors, fragrances, pharmaca and pheromones. Weinheim: Wiley-VCH; 2006.
Bucchi G, Wuest H. New synthesis of agarofuran and of dihydroagarofuran. J Org Chem. 1979;44:44–5.
Burfield T, Reekie SL. Mosquitoes, malaria and essential oils. Int J Aromather. 2005;15(1):30–41.
Burfield T, Kirkham K. The agarwood files, the Cropwatch (online). 2005. Retrieved from http://www.cropwatch.org/agarwood.htm.
Burt S. Essential oils: Their antibacterial properties and potential applications in foods – a review. Int J Food Microbiol. 2004;94(3):223–53.
Cardeal ZL, Marriott PJ. Comprehensive two-dimensional gas chromatography-mass spectrometry, analysis and comparison of volatile organic compounds in Brazilian *cachaca* and selected spirits. Food Chem. 2009;112:747–55.
Chang YS, Nor Azah MA, Abu Said A, Lok EH, Reader S, Spiers A. Gaharu. FRIM Tech Inf. 2002;69:1–7.
Chen H, Yang Y, Xue J, Wei J, Zhang Z, Chen H. Comparison of compositions and antimicrobial activities of essential oils from chemically stimulated agarwood, wild agarwood and healthy *Aquilaria sinensis* (Lour.) Gilg trees. Molecules. 2011;16:4884–96.
Chen HQ, Wei JH, Yang JS, Zhang Z, Yang Y, Gao ZH, Sui C, Gong B. Chemical constituents of agarwood originating from the endemic genus *Aquilaria* plants. Chem Biodivers. 2012;9:236–49.
Dimandja JM, Stanfill S, Grainger J. Application of comprehensive two-dimensional gas chromatography to the qualitative analysis of essential oils. J High Resolut Chromatogr. 2000;23:208–14.
Edris AE. Pharmaceutical and therapeutic potentials of essential oils and their individual volatile constituents: A review. Phytother Res. 2007;21(4):308–23.
Ernest G. The essential oils, History-origin in plants-production-analysis, vol. 1. London: Lancaster Press Inc; 1948. p. 1–426.
Giddings JC. Two-dimensional separations: concept and promise. Anal Chem. 1984;56(12):258–1270.
Gross JH. Mass spectrometry: a textbook. Berlin: Springer; 2004.
Ishihara M, Tsuneya T. Components of the volatile concentrate of agarwood. J Essen Oil Res. 1993a;5:283–9.
Ishihara M, Tsuneya T. Components of the agarwood smoke on heating. J Essen Oil Res. 1993b;5:419–23.
Ishihara M, Tsuneya T, Suga M, Uneyama K. Three sesquiterpenes from agarwood. Phytochemistry. 1991a;30(10):563–6.
Ishihara M, Tsuneya T, Suga M, Uneyama K. Guaiane sesquiterpenes from agarwood. Phytochemistry. 1991b;30(10):3343–7.
Jayachandran K, Sekar I, Parthiban KT, Amirtham D, Suresh KK. Analysis of difference grades of agarwood (*Aquilaria malaccensis* Lamk.) oil through GC-MS. Indian J Nat Prod Resour. 2014;5(1):44–7.

Joulain D, König WA. The atlas of spectral data of sesquiterpene hydrocarbons. Hamburg: E.B.-Verlag; 1998.

Khummueng W, Trenerry C, Rose G, Marriot PJ. Application of comprehensive two-dimensional gas chromatography with nitrogen-selective detection for the analysis of fungicide residues in vegetable samples. J Chromatogr A. 2006;1131:203–14.

König WA, Bulow N, Saritas Y. Identification of sesquiterpene hydrocarbons by gas phase analytical methods. Flavour Fragrance J. 1999;14:367–78.

Luque de Castro MD, Jiménez-Carmona MM, Fernández-Pérez V. Towards more rational techniques for the isolation of valuable essential oils from plants. TrAC Trends Anal Chem. 1999;18(11):708–16.

Ma C, Wang H, Lu X, Li H, Liu B, Xu G. Analysis of *Artemisia annua* L. volatile oil by comprehensive two-dimensional gas chromatography time-of-flight mass spectroscopy. J Chromatogr A. 2007;1150:50–3.

Maheshwari ML, Jain TC, Bates RB, Bhattacharyya SC. Terpenoids: structure and absolute configuration of α-agarofuran, β-agarofuran and dihydroagarofuran. Tetrahedron. 1963;19: 1079–90.

Marriott PJ, Shellie R. Principles and applications of comprehensive two-dimensional gas chromatography. TrAC Trends Anal Chem. 2002;21(9-10):573–83.

Marriott PJ, Shellie R, Cornwell C. Gas chromatographic technologies for the analysis of essential oils. J Chromatogr A. 2001;936(1-2):1–22.

Micyus NJ, McCurry JD, Seeley JV. Analysis of aromatic compound in gasoline with flow-switching comprehensive two-dimensional gas Chromatography. J Chromatogr A. 2005;1086(1-2):115–21.

Min W, Marriott PJ, Chan WH, Lee AWM, Lee AW, Huei CW. Enantiomeric separation and quantification of ephedrine-type alkaloids in herbal materials by comprehensive two-dimensional gas chromatography. J Chromatogr A. 2006;1112:361–8.

Mondello L, Casilli A, Tranchida PQ, Dugo P, Dugo G. Comprehensive two-dimensional gas chromatography for the analysis of citrus essential oils. Flavour Fragrance J. 2005;20(2):136–40.

Mondello L, Tranchida PQ, Dugo P, Dugo G. Comprehensive two-dimensional gas chromatography-mass spectrometry: a review. Mass Spectrom Rev. 2007;27:101–24.

Naef R. The volatile and semi-volatile constituents of agarwood, the infected heartwood of *Aquilaria* species: a review. Flavour Fragrance J. 2011;26:73–89.

Nakanishi T, Yamagata E, Yoneda K, Miura I. Jinkohol, a prezizane sesquiterpene alcohol from agarwood. Phytochemistry. 1981;20:1597–9.

Nakanishi T, Yamagata E, Yoneda K, Miura I, Mori H. Jinkoh-eremol and jinkohol II, two sesquiterpene alcohols from agarwood. J Chem Soc Perkin Trans. 1983;1:601–4.

Nakanishi T, Yamagata E, Yoneda K, Nagashima T, Kawasaki I, Yoshida T, Mori H, Miura I. Three fragrant sesquiterpenes of agarwood. Phytochemistry. 1984;22:2066–7.

Nor Azah MA, Chang YS, Mailina J, Abu Said A, Abd Majid J, Saidatul Husni S, Nor Hasnida H, Nik Yasmin Y. Comparison of chemical profiles of selected gaharu oils from Peninsular Malaysia. Malays J Anal Sci. 2008;12(2):338–40.

Pornpunyapat J, Chetpattananondh P, Tongurai C. Mathematical modeling for extraction of essential oil from *Aquilaria crassna* by hydrodistillation and quality of agarwood oil. Bangladesh J Pharmacol. 2011;6:18–24.

Pripdeevech P, Wongpornchai S, Marriott PJ. Comprehensive two-dimensional gas chromatography-mass spectrometry analysis of volatile constituents in Thai vetiver root oils obtained by using different extraction methods. Phytochem Anal. 2010;21(2):163–73.

Rocha SM, Coelho E, Zrostlikova J, Delgadillo I, Coimbra MA. Comprehensive two-dimensional gas chromatography with time-of-flight mass spectrometry of monoterpenoids as a powerful tool for grape origin traceability. J Chromatogr A. 2007;1161:292–9.

Schoenmakers PJ, Oomen JL, Blomberg J, Genuit W, Velzen GV. Comparison of comprehensive two-dimensional gas chromatography and gas chromatography-mass spectrometry for the characterization of complex hydrocarbon mixtures. J Chromatogr A. 2000; 892(1-2):29–46.

Shellie R, Marriott PJ, Cornwell C. Application of comprehensive two-dimensional gas chromatography (GC×GC) to the enantioselective analysis of essential oils. J Sep Sci. 2001;24: 823–30.

Stewart D. The chemistry of essential oils made simple: God's love manifest in molecules. Marble Hill: Care Publication; 2005.

Tajuddin SN. Identification of sesquiterpenes hydrocarbon compounds in volatile agarwood oils using gas phase methods. Doctoral dissertation, University of Malaysia Pahang, Gambang; 2010.

Tajuddin SN, Yusoff MM. Chemical composition of volatile oils of *Aquilaria malaccensis* (Thymelaeaceae) from Malaysia. Nat Prod Commun. 2010;5(12):1965–8.

Terry A, Heinrich A. Flavornet and human odor space, gas chromatography – olfactometry (GCO) of natural products. 1984. Retrieved from http://www.flavornet.org/.

Varma KR, Maheshwari ML, Bhattacharyya SC. Terpenoids-LXII: the constitution of agarospirol, a sesquiterpenoid with a new skeleton. Tetrahedron. 1965;21:115–38.

Wetwitayaklung P, Thavanapong N, Charoenteeraboon J. Chemical constituents and antimicrobial activity of essential oil and extracts of heartwood of *Aquilaria crassna* obtained from water distillation and supercritical fluid carbon dioxide extraction. Silpakorn University Sci Technol J. 2009;3:25–33.

Wu J, Lu X, Tang W, Kong H, Zhou S, Xu G. Application of comprehensive two-dimensional gas chromatography-time-of-flight mass spectrometry in the analysis of volatile oil of traditional Chinese medicines. J Chromatogr A. 2004;1034:199–205.

Xing LZ, Yang YL, Jian HW, Yun Y, Zheng Z, Jun QH, Huai QC, Yu JL. Production of high quality agarwood in *Aquilaria sinensis* trees via whole-tree agarwood-induction technology. Chin Chem Lett. 2012;23:727–30.

Yeung AY. Encyclopedia of common natural ingredients. New York: John Wiley & Sons; 1980.

Yoneda K, Yamagata E, Nakanishi T, Nagashima T, Kawasaki I, Yoshida T, Mori H, Miura I. Sesquiterpenoids in two different kinds of agarwood. Phytochemistry. 1984;23:2068–9.

Yoswathana N. Extraction of agarwood (*Aquilaria crassna*) oil by using supercritical carbon dioxide extraction and enzyme pretreatment on hydrodistillation. J Food Agric Environ. 2013;11(2):1055–9.

Yoswathana N, Eshiaghi MN, Jaturapornpanich K. Enhancement of essential oil from agarwood by subcritical water extraction and pretreatments on hydrodistillation. World Acad Sci Eng Technol. 2012;65:869–75.

Chapter 8
Pharmacological Effects of *Aquilaria* spp. Leaves and Their Chemical Constituents

Mamoru Kakino and Hideaki Hara

Abstract *Aquilaria* is one of the most famous and historical botanical tree species in Southeast Asia, the Middle East, and the Far East including Japan. Resin-containing wood from *Aquilaria* trees is known as agarwood or *gaharu*. Many studies have been conducted on the wood of *Aquilaria*; however, it is not the case with the leaves. *Aquilaria* leaves have been proclaimed as effective for the treatment of several diseases, although few cases had been reported until 2008. We have previously reported that compounds from *Aquilaria* had a laxative effect in a mouse model of constipation, and the main active constituents include mangiferin and genkwanin-5-*O*-*beta*-premeveroside. In this chapter, we review the biological effects of *Aquilaria* species and its main chemical constituents, especially mangiferin, which is one of the active constituents that shows a laxative effect.

8.1 Introduction

Aquilaria leaves are rich in different compounds, many of which have not been tapped. Some of these compounds possess important pharmacological bioactivities that are beneficial to human. We have studied compounds from *Aquilaria sinensis* leaves (Fig. 8.1) extracted with acetone, which significantly increased stool volume in normal mice (Hara et al. 2008). We also found that not only *A. sinensis* but also *A. crassna* has a laxative effect in constipation-induced mouse. The main active constituents of both species include mangiferin and genkwanin-5-*O*-*beta*-premeveroside (Kakino et al. 2010a).

In recent years, many pharmacological activities of *Aquilaria* spp. other than the laxative effects have been reported, such as antibacterial activity (Kamonwannasit et al. 2013), inhibition of nitric oxide production (Yang et al. 2012), antihyperglycemic activity in *db*/*db* mice (Jiang et al. 2011), antinociceptive effect (Zhou et al. 2008), alpha-glucosidase inhibitory activity (Feng et al. 2011), analgesic activity (Sattayasai

M. Kakino • H. Hara (✉)
Molecular Pharmacology, Department of Biofunctional Evaluation,
Gifu Pharmaceutical University, Gifu 501-1196, Japan
e-mail: hidehara@gifu-pu.ac.jp

R. Mohamed (ed.), *Agarwood*, Tropical Forestry,
DOI 10.1007/978-981-10-0833-7_8

Fig. 8.1 Leaves of *Aquilaria sinensis* collected in Taiwan

et al. 2012), and acetylcholinesterase inhibitory activity (Bahrani et al. 2014). To confirm the laxative effect of compounds from these tree species, we performed five trials in humans and found that an extract of *Aquilaria crassna* had a laxative effect in individuals with a tendency toward constipation.

The chemical constituents of *Aquilaria* spp. have been reported in many studies (Nie et al. 2009; Qi et al. 2009; Feng et al. 2011; Feng and Yang 2011; Feng and Yang 2012; Ito et al. 2012; Cheng et al. 2013; Xia et al. 2013; Yu et al. 2013; Tay et al. 2014). Of the constituents studied, mangiferin and epigallocatechin gallate (EGCG) have been discussed the most. The biological activities of most polyphenolic glycosides have not been reported as much because constituents such as genkwanin-5-*O*-*beta*-premeveroside are specific to *Aquilaria* species.

8.2 Effects of *Aquilaria* Species

8.2.1 Effect on the Gastrointestinal Tract

8.2.1.1 Laxative Effect

Aquilaria crassna and *A. sinensis* have been reported to have a laxative effect in mouse model of loperamide-induced constipation. *Aquilaria sinensis* has also been

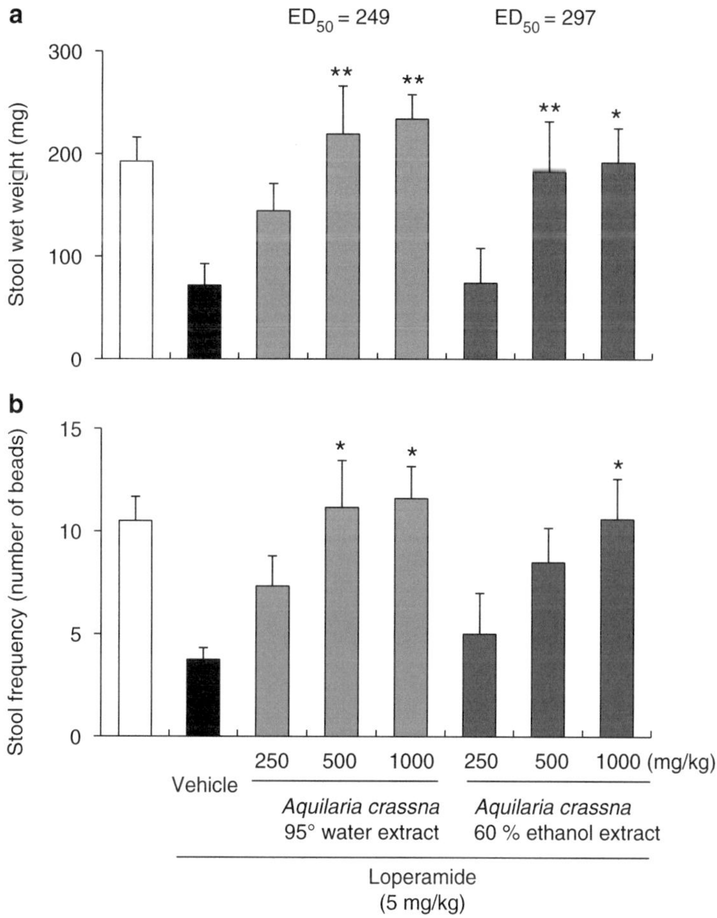

Fig. 8.2 Laxative effect of *Aquilaria crassna* leaf extracts (a water extract with a temperature of 95 °C and a 60 % ethanol extract) in mice with loperamide-induced constipation (n = 4–6). Stool wet weight (**a**). Stool frequency (**b**). $*p < 0.05$, $**p < 0.01$ vs. vehicle (one-way ANOVA and Holm's multiple comparison test)

reported to have a laxative effect in a rat model of low-fiber-induced constipation (Kakino et al. 2010a, b). In these studies, the ethanol extract of *Aquilaria* spp. leaves significantly increased the number and weight of stool beads in each experimental animal. Further, a 60 % ethanol extract of *A. sinensis* has been reported to increase the contraction tension of the isolated jejunum and ileum (Kakino M, 2012, PhD Thesis, Gifu Pharmaceutical University, Gifu, Japan). In fact, extracts of both *A. sinensis* and *A. crassna* increase the contraction tension of the small intestine. In addition to the 60 % ethanol extract of *A. crassna*, its water extract with a temperature of 95 °C has been reported to have a laxative effect and increase the contraction tension of the intestines (Figs. 8.2 and 8.3, Kakino M, 2012, PhD Thesis, Gifu Pharmaceutical University, Gifu, Japan).

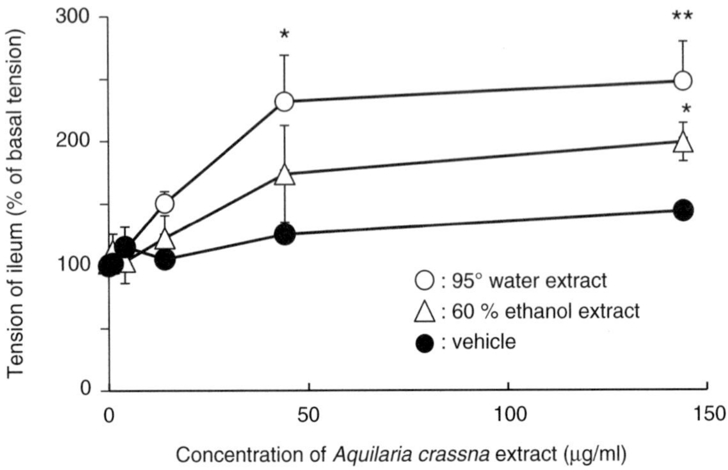

Fig. 8.3 Effect of *Aquilaria crassna* extract on ileum tension. The 95 °C water extract and 60 % ethanol extract were administered at concentrations of 1, 3, 10, 30, and 100 μg/ml cumulatively. DMSO was administered at 0.003, 0.01, 0.03, and 0.1 % (v/v) cumulatively. Tension was shown as percent of the tension before sample administration. Data are shown as mean ± SEM. $**p < 0.01$, $**p < 0.05$ vs. vehicle ($n = 3$, paired Tukey's multiple comparison test). ○ water extract with a temperature of 95 °C, ● DMSO, △ 60 % ethanol extract

8.2.1.2 Decrease in Intestinal Toxins via Antimicrobial Activity

Apart from the laxative effect, *A. crassna* also decreases the intestinal toxins, ammonium, and indole, in high-protein high-fat diet-induced putrefaction mice model. Both the water extract (95 °C) and 60 % ethanol extract of *A. crassna* decreased the indole in feces, while only the water extract was found to decrease ammonium in feces (Kakino et al. 2012). *Aquilaria crassna* also exhibited antimicrobial activities against *Bacteroides vulgatus*, *Bacteroides fragilis*, *Staphylococcus aureus*, *Clostridium difficile*, and *Peptostreptococcus anaerobius* with minimum inhibitory concentration (MIC) values of 4–8 mg/ml in vitro (Table 8.1). However, *Aquilaria crassna* had no effect on *Escherichia coli*, *Enterococcus faecalis*, *Bifidobacterium longum*, and *Bifidobacterium adolescentis* with MIC values less than 8 mg/ml (Kakino et al. 2012). *Aquilaria crassna* leaves are also reported to have antimicrobial activities against *Staphylococcus epidermidis* in vitro (Kamonwannasit et al. 2013). The decrease in the intestinal toxins, ammonium, and indole in feces can be partly attributed to its antimicrobial activities on the intestinal microbes.

With regard to intestinal toxins, urease activities in the gastrointestinal tract have a large influence on ammonium genesis. *Daphne retusa*, which does not belong to *Aquilaria* but a member of the Thymelaeaceae family, contains urease inhibitory constituents (Mansoor et al. 2014). Though it has not been reported yet, further study is needed to reveal whether *Aquilaria* spp. as well as *D. retusa* contain urease inhibitory constituents.

Table 8.1 Antimicrobial effect of *Aquilaria crassna* extract expressed as minimum inhibitory concentration (MIC) (Kakino et al. 2012)

	MICs of agarwood extracts (mg/ml)		
	WEA	EEA	Ammoniagenesis
G (–) bacteria			
Escherichia coli	>8	>8	+
Bacteroides vulgatus	8	8	+
Bacteroides fragilis	8	8	+
G (+) bacteria			
Staphylococcus aureus	4	4	++
Clostridium difficile	8	4	+
Peptostreptococcus anaerobius	4	4	++
Enterococcus faecalis	>8	>8	+
Bifidobacterium longum	>8	>8	–
Bifidobacterium adolescentis	>8	>8	–

WEA hot water extract with a temperature of 95 °C, *EEA* 60 % ethanol extract

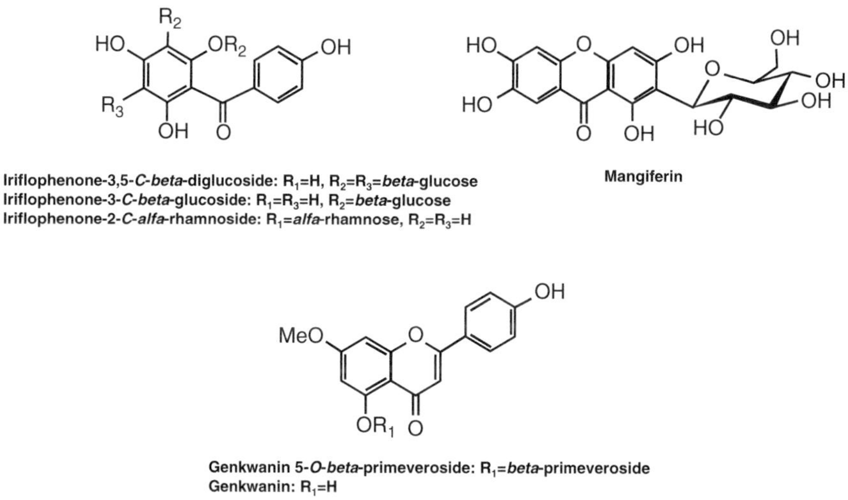

Fig. 8.4 Main chemical constituents in *Aquilaria* leaves identified using HPLC (330 nm)

8.2.1.3 Alpha-Glucosidase Inhibitory Activity

Eight alpha-glucosidase inhibitors, including four new compounds, were isolated from the 70 % ethanol extract of *A. sinensis*: aquilarisinin, aquilarisin, hypolaetin 5-*O*-*beta*-D-glucuronopyranoside, aquilarixanthone, mangiferin, iriflophenone 2-*O*-*alpha*-L-rhamnopyranoside, iriflophenone 3-C-*beta*-D-glucoside, and iriflophenone 3,5-C-*beta*-D-diglucopyranoside (Feng et al. 2011) (Fig. 8.4). The ethyl acetate fraction of *Thymelaea hirsute*, which also belongs to the Thymelaeaceae family, has demonstrated alpha-glucosidase inhibitory activity in vitro and a decrease in intestinal glucose uptake in vivo. Consequently, *T. hirsute* decreases the postprandial hyperglycemia in normal and diabetic mice (Abid et al. 2014). Taken together, some Thymelaeaceae species as well as *A. sinensis* and *T. hirsute* may contain alpha-glucosidase inhibitory chemical compounds.

8.2.2 *Pharmacological Effects of* Aquilaria *spp. Derived from Antioxidative Activity*

8.2.2.1 Antipyretic, Analgesic, Antioxidative, and Anti-inflammatory Activities

There are several reports that discuss the biological activities derived from the antioxidative activities (Zhou et al. 2008; Chen et al. 2011; Sattayasai et al. 2012; Yang et al. 2012). *Aquilaria crassna* is reported to have antipyretic activities in Baker's yeast-induced fever in rats, analgesic activities in hot plate test in mice, anti-inflammatory activities in carrageenan-induced paw edema in rats, and anti-oxidative activities by 2,2-diphenyl-1-picrylhydrazyl (DPPH) assay (Sattayasai et al. 2012). *Aquilaria sinensis* is also reported to have anti-inflammatory activity in xylene-induced edema model, carrageenan-induced edema model, and carboxymethylcellulose sodium-induced leukocyte migration model in mice (Zhou et al. 2008). *Aquilaria sinensis* was also reported to have inhibitory effect on nitric oxide (NO) release in lipopolysaccharide (LPS)-induced mouse peritoneal macrophages model in vitro (Zhou et al. 2008). Recently, *A. sinensis* was also reported to have an inhibitory effect on NO release in LPS-induced RAW 264.7 macrophages model (Yang et al. 2012). Various biological effects of *A. sinensis* leaves cultivated under different conditions were compared to evaluate the differences in the activity of enzymes such as peroxidase, catalase, and superoxide dismutase (Chen et al. 2011).

Apart from *Aquilaria* spp., *Daphne pontica*, which belongs to the Thymelaeaceae family, is reported to have anti-inflammatory activities in carrageenan-induced edema model, prostaglandin E2-induced edema model, and 12-O-tetradecanoyl-13-acetate-induced edema model in mice (Kupeli et al. 2007). On the other hand, *D. pontica* is not reported to show significant antinociceptive activity (Kupeli et al. 2007).

8.2.3 *AMP-Activated Protein Kinase (AMPK)-Activating Effect*

Aquilaria sinensis is reported to reduce fasting blood glucose and glycosylated hemoglobin levels in db/db mice with ameliorating effect on insulin resistance (Jiang et al. 2011). In this model, *A. sinensis* is reported to have a thiazolidinedione (TZD)-like activity to activate AMPK (Jiang et al. 2011).

Interestingly, we already mentioned that *A. sinensis* contains alpha-glucosidase inhibitory constituents (Feng et al. 2011). The biological effect on db/db mice is hypothesized to be related to both alpha-glucosidase inhibitory activity and AMPK-activating effect.

8.2.4 Acetylcholinesterase (AChE) Inhibitory Activity

Aquilaria subintegra leaves have been claimed to be effective for the treatment of Alzheimer's disease by a traditional practitioner in Malaysia. The crude extract of *A. subintegra* and its chemical constituents, kaempferol 3,4,7-trimethyl ether, are reported to have AChE inhibitory effects (Bahrani et al. 2014). The crude extract of *A. subintegra*, kaempferol, and 3,4,7-trimethyl ether significantly reduced the number of repeated entries into the arms of radial arm maze (RAM) in valium-impaired memory model.

8.3 Effect of Chemical Constituents of *Aquilaria*

8.3.1 Polyphenolic Glycosides

We reported that both ethanol extract and water extract (95 °C) of *A. crassna* contain mangiferin, iriflophenone-3,5-C-*beta*-diglucoside, iriflophenone-3-C-*beta*-glucoside, iriflophenone-2-C-*alfa*-rhamnoside, genkwanin, and genkwanin 5-*O*-*beta*-primeveroside (Fig. 8.3, unpublished data of authors). Among these compounds, mangiferin and genkwanin 5-*O*-*beta*-primeveroside are reported to have laxative effect (Kakino et al. 2010a). On the other hand, as already mentioned, mangiferin, iriflophenone-3,5-C-*beta*-diglucoside, and iriflophenone-3-C-*beta*-glucoside are reported to have alpha-glucosidase inhibitory activity (Feng et al. 2011).

8.3.2 Mangiferin

Among the chemical constituents isolated from *Aquilaria* species, mangiferin is most widely reported. Theoretically, *Aquilaria* species should have the same pharmacological activities as mangiferin. Hence, we discuss here the pharmacological activities of mangiferin.

8.3.2.1 Inhibitory Activities Against Several Carbohydrate-Metabolizing Enzymes Including Alpha-Glucosidase

A. sinensis is reported to have alpha-glucosidase inhibitory activity (Feng et al. 2011). Mangiferin is also reported to have inhibitory effect against several carbohydrates, including alpha-glucosidase, sucrase, maltase, isomaltase, alpha-amylase, and aldose reductase (Yoshikawa et al. 2001).

8.3.2.2 AMPK-Activating Effect

As mentioned earlier, *A. sinensis* is reported to activate AMPK in order to reduce the fasting blood glucose and glycosylated hemoglobin levels in db/db mice, consequently resulting in an ameliorating effect on insulin resistance (Jiang et al. 2011). Mangiferin is reported to activate AMPK in L6 myotubes in vitro (Wang et al. 2014).

8.3.2.3 Antidiabetic Effect in KK-A^y Diabetic Model, Streptozotocin-Induced Diabetic Mice, and High-Fat Diet-Induced Obesity Mice

An insulin tolerance test showed that mangiferin and its glucoside ameliorated blood glucose levels in the KK-A^y spontaneous diabetes model mice (Miura et al. 2001). In conventional rats, pseudo-germ-free rats, and streptozotocin (STZ)-induced diabetic rats, orally administered mangiferin was detected in urine, blood, and feces (Liu et al. 2012). In streptozotocin-induced diabetic rats, mangiferin significantly ($p < 0.05$) lowered the level of blood glucose and altered the levels of biochemical parameters including urea, uric acid, and creatinine. In addition, red blood cell and white blood cell counts and their functional indices were significantly improved and toxicological parameters including AST, ALT, and ALP levels were also significantly reduced (Sellamuthu et al. 2014). In the same rat diabetic model, mangiferin ameliorated blood glucose and increased plasma insulin levels. The activities of antioxidant enzymes, such as superoxide dismutase, catalase, and glutathione peroxidase, and the level of reduced glutathione were significantly decreased, suggesting that glucose toxicity in blood decreased because of mangiferin administration (Sellamuthu et al. 2013). Thus, mangiferin attenuated renal fibrosis via suppression of osteopontin overproduction (Zhu et al. 2014).

In high-fat diet-induced obesity mice, mangiferin attenuated hyperglycemia, insulin resistance, and hyperlipidemia. Simultaneously, it significantly increased the levels of proteins important for mitochondrial biogenesis and oxidative activity, including oxoglutarate dehydrogenase E1 and cytochrome c oxidase subunit 6B1, and decreased the levels of those critical for lipogenesis, such as fatty acid stearoyl-CoA desaturase 1 and acetyl-CoA carboxylase 1 (Lim et al. 2014).

8.3.2.4 Accelerating Activity on Gastrointestinal Contents

As we previously reported, *A. crassna* and *A. sinensis* have a laxative effect in mice with loperamide-induced constipation and mice and rats with low-fiber-induced constipation (Kakino et al. 2010a, 2012). In both models, extracts of *A. sinensis* and *A. crassna* significantly accelerated gastrointestinal contents. With regard to mangiferin, we confirmed the laxative effect but did not confirm the accelerating effect on gastrointestinal contents. Another group reported that mangiferin significantly accelerated gastrointestinal contents in normal mice and in mice with morphine-induced constipation (Cavalcante Morais et al. 2012).

8.3.2.5 Antidepressive and Antianxiety Effects

Mangiferin pretreatment is reported to have antidepressant and antianxiety effects via reduction of interleukin-1 beta levels and oxidative stress induced by LPS (Jangra et al. 2014). Mangiferin-containing *Aquilaria* spp. are also expected to be useful for the treatment of depressive and anxiety illnesses. In this chapter, the anti-oxidative activity of mangiferin is suggested to protect against LPS-induced oxidative stress.

8.3.2.6 Anti-oxidative Stress Effect, Ameliorating Effect Against DNA Damage, and Anti-inflammatory Effect

Mangiferin is reported to ameliorate chemical carcinogen-induced DNA damage via activation of the Nrf2-ARE pathway, which is a redox-sensitive pathway, in etoposide-treated mononuclear cells (Zhang et al. 2015). In another model, mangiferin inhibited Nrf2 ubiquitination in HL60 cells without increasing the expression of Nrf2 mRNA (Zhang et al. 2014; Zhao et al. 2014). With regard to an animal model, mangiferin inhibited the production of proinflammatory mediators via inhibition of sepsis-activated mitogen-activated protein kinases and nuclear factor kappa-light-chain-enhancer of activated B cell signaling, resulting in attenuation of cecal ligation and puncture-induced mortality and acute lung injury in mice (Gong et al. 2013).

8.4 Conclusion and Future Perspectives

Many pharmacological effects of *Aquilaria* spp. leaves have been reported since 2008, when our group published the first one. The summary of the effects is shown (Fig. 8.5). Among these effects, only laxative effect is published in the human

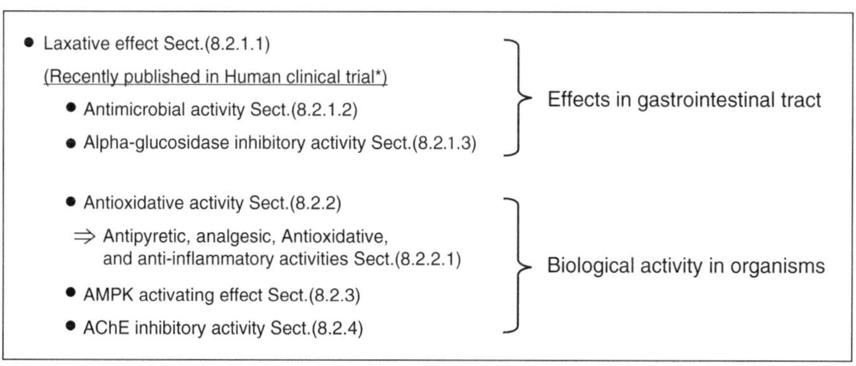

Fig. 8.5 Pharmacological effects of *Aquilaria* leaves, published up to Jan 2015 (Kakino et al. 2015)

clinical trial ensuring laxative effect and clinical safety. The published activities of *Aquilaria* spp. leaves such as the AMPK-activating activity and AchE inhibitory activity have yet to be proven in animal models. We hope to conduct these in future studies. While writing this manuscript, we had searched the world's most popular database, PubMed (http://www.ncbi.nlm.nih.gov/pubmed/), using "Aquilaria" as the keyword. Pleasantly surprised, we found that over 50 % of the reported studies were made in the last five years (2010–2015) and over 75 % in the last ten years. This tendency indicates that the worldwide research on *Aquilaria* spp. leaves from the viewpoint of biochemistry, pharmacology, and medical science has been accelerating in the recent years.

References

Abid S, Lekchiri A, Mekhfi H, Ziyyat A, Legssyer A, Aziz M, et al. Inhibition of alpha-glucosidase and glucose intestinal absorption by *Thymelaea hirsuta* fractions. J Diabetes. 2014;6(4):351–9. Epub 2013/11/14.

Bahrani H, Mohamad J, Paydar MJ, Rothan HA. Isolation and characterisation of acetylcholinesterase inhibitors from *Aquilaria subintegra* for the treatment of Alzheimer's disease (AD). Curr Alzheimer Res. 2014;11(2):206–14. Epub 2014/02/01.

Cavalcante Morais T, Cavalcante Lopes S, Bezerra Carvalho KM, Rodrigues Arruda B, Correia de Souza FT, Salles Trevisan MT, et al. Mangiferin, a natural xanthone, accelerates gastrointestinal transit in mice involving cholinergic mechanism. World J Gastroenterol WJG. 2012; 18(25):3207–14. Epub 2012/07/12.

Chen JY, Xiong Y, Deng MH. Study on the physiological and biochemical characteristics of different types of leaves of *Aquilaria sinensis*. Zhong yao cai J Chinese Med Mater. 2011;34(11):1664–6. Epub 2012/04/18.

Cheng JT, Han YQ, He J, De Wu X, Dong LB, Peng LY, et al. Two new tirucallane triterpenoids from the leaves of *Aquilaria sinensis*. Arch Pharm Res. 2013;36(9):1084–9. Epub 2013/03/20.

Feng J, Yang X. Liposolubility constituents from leaves of *Aquilaria sinensis*. Zhongguo zhongyao zazhi China J Chinese Mater Med. 2011;36(15):2092–5. Epub 2011/11/10.

Feng J, Yang X. Constituents from the leaves of *Aquilaria sinensis*. Zhongguo zhongyao zazhi China J Chinese Mater Med. 2012;37(2):230–4. Epub 2012/06/29.

Feng J, Yang XW, Wang RF. Bio-assay guided isolation and identification of alpha-glucosidase inhibitors from the leaves of *Aquilaria sinensis*. Phytochemistry. 2011;72(2–3):242–7. Epub 2011/01/11.

Gong X, Zhang L, Jiang R, Ye M, Yin X, Wan J. Anti-inflammatory effects of mangiferin on sepsis-induced lung injury in mice via up-regulation of heme oxygenase-1. J Nutr Biochem. 2013;24(6):1173–81. Epub 2012/12/26.

Hara H, Ise Y, Morimoto N, Shimazawa M, Ichihashi K, Ohyama M, et al. Laxative effect of agarwood leaves and its mechanism. Biosci Biotechnol Biochem. 2008;72(2):335–45. Epub 2008/02/08.

Ito T, Kakino M, Tazawa S, Watarai T, Oyama M, Maruyama H, et al. Quantification of polyphenols and pharmacological analysis of water and ethanol-based extracts of cultivated agarwood leaves. J Nutr Sci Vitaminol. 2012;58(2):136–42. Epub 2012/07/14.

Jangra A, Lukhi MM, Sulakhiya K, Baruah CC, Lahkar M. Protective effect of mangiferin against lipopolysaccharide-induced depressive and anxiety-like behaviour in mice. Eur J Pharmacol. 2014;740:337–45. Epub 2014/07/30.

Jiang S, Jiang Y, Guan Y-F, Tu P-F, Wang K-Y, Chen J-M. Effects of 95% ethanol extract of *Aquilaria sinensis* leaves on hyperglycemia in diabetic db/db mice. J Chinese Pharm Sci. 2011;20(6):609–14.

Kakino M, Tazawa S, Maruyama H, Araki Y, Ichihara K, Kouno H, Iinuma M, Hara H. A double-blinded, placebocontrolled, crossover trial and open-label safety trials of extract of *Aquilaria* sp Thymelaeaceae leaves. Jpn Pharmacol Ther. 2015;43(2):209–29.

Kakino M, Izuta H, Ito T, Tsuruma K, Araki Y, Shimazawa M, et al. Agarwood induced laxative effects via acetylcholine receptors on loperamide-induced constipation in mice. Biosci Biotechnol Biochem. 2010a;74(8):1550–5. Epub 2010/08/12.

Kakino M, Tazawa S, Maruyama H, Tsuruma K, Araki Y, Shimazawa M, et al. Laxative effects of agarwood on low-fiber diet-induced constipation in rats. BMC Complement Altern Med. 2010b;10:68. Epub 2010/11/17.

Kakino M, Sugiyama T, Kunieda H, Tazawa S, Maruyama H, Tsuruma K, et al. Agarwood (*Aquilaria crassna*) extracts decrease high-protein high-fat diet-induced intestinal putrefaction toxins in mice. Pharm Anal Acta. 2012;3:152.

Kamonwannasit S, Nantapong N, Kumkrai P, Luecha P, Kupittayanant S, Chudapongse N. Antibacterial activity of *Aquilaria crassna* leaf extract against *Staphylococcus epidermis* by disruption of cell wall. Ann Clin Microbiol Antimicrob. 2013;12:20. Epub 2013/08/22.

Kupeli E, Tosun A, Yesilada E. Assessment of anti-inflammatory and antinociceptive activities of *Daphne pontica* L. (Thymelaeaceae). J Ethnopharmacol. 2007;113(2):332–7. Epub 2007/08/08.

Lim J, Liu Z, Apontes P, Feng D, Pessin JE, Sauve AA, et al. Dual mode action of mangiferin in mouse liver under high fat diet. PLoS One. 2014;9(3):e90137. Epub 2014/03/07.

Liu H, Wu B, Pan G, He L, Li Z, Fan M, et al. Metabolism and pharmacokinetics of mangiferin in conventional rats, pseudo-germ-free rats, and streptozotocin-induced diabetic rats. Drug Metab Dispos Biol Fate Chem. 2012;40(11):2109–18. Epub 2012/08/04.

Mansoor F, Anis I, Khan A, Marasini BP, Choudhary MI, Shah MR. Urease inhibitory constituents from *Daphne retusa*. J Asian Nat Prod Res. 2014;16(2):210–5. Epub 2013/11/26.

Miura T, Ichiki H, Iwamoto N, Kato M, Kubo M, Sasaki H, et al. Antidiabetic activity of the rhizoma of *Anemarrhena asphodeloides* and active components, mangiferin and its glucoside. Biol Pharm Bull. 2001;24(9):1009–11. Epub 2001/09/18.

Nie C, Song Y, Chen D, Xue P, Tu P, Wang K, et al. Studies on chemical constituents of leaves of *Aquilaria sinensis*. Zhongguo zhongyao zazhi China J Chinese Mater Med. 2009;34(7):858–60. Epub 2009/07/25.

Qi J, Lu JJ, Liu JH, Yu BY. Flavonoid and a rare benzophenone glycoside from the leaves of *Aquilaria sinensis*. Chem Pharm Bull. 2009;57(2):134–7. Epub 2009/02/03.

Sattayasai J, Bantadkit J, Aromdee C, Lattmann E, Airarat W. Antipyretic, analgesic and anti-oxidative activities of *Aquilaria crassna* leaves extract in rodents. J Ayurveda Integr Med. 2012;3(4):175–9. Epub 2013/01/18.

Sellamuthu PS, Arulselvan P, Kamalraj S, Fakurazi S, Kandasamy M. Protective nature of mangiferin on oxidative stress and antioxidant status in tissues of streptozotocin-induced diabetic rats. ISRN Pharm. 2013;2013:750109. Epub 2013/10/30.

Sellamuthu PS, Arulselvan P, Fakurazi S, Kandasamy M. Beneficial effects of mangiferin isolated from *Salacia chinensis* on biochemical and hematological parameters in rats with streptozotocin-induced diabetes. Pak J Pharm Sci. 2014;27(1):161–7. Epub 2014/01/01.

Tay PY, Tan CP, Abas F, Yim HS, Ho CW. Assessment of extraction parameters on antioxidant capacity, polyphenol content, epigallocatechin gallate (EGCG), epicatechin gallate (ECG) and iriflophenone 3-C-beta-glucoside of agarwood (*Aquilaria crassna*) young leaves. Molecules. 2014;19(8):12304–19. Epub 2014/08/26.

Wang F, Yan J, Niu Y, Li Y, Lin H, Liu X, et al. Mangiferin and its aglycone, norathyriol, improve glucose metabolism by activation of AMP-activated protein kinase. Pharm Biol. 2014;52(1):68–73. Epub 2013/09/17.

Xia F, Sun J, Jiang Y, Tu PF. Further chemical investigation of leaves of *Aquilaria sinensis*. Zhongguo Zhong yao za zhi 2013;38(19):3299–303. Epub 2014/01/16.

Yang XB, Feng J, Yang XW, Zhao B, Liu JX. Aquisiflavoside, a new nitric oxide production inhibitor from the leaves of *Aquilaria sinensis*. J Asian Nat Prod Res. 2012;14(9):867–72. Epub 2012/08/29.

Yoshikawa M, Nishida N, Shimoda H, Takada M, Kawahara Y, Matsuda H. Polyphenol constituents from *Salacia* species: quantitative analysis of mangiferin with alpha-glucosidase and aldose reductase inhibitory activities. Yakugaku zasshi J Pharm Soc Jpn. 2001;121(5):371–8. Epub 2001/05/22.

Yu Q, Qi J, Yu HX, Chen LL, Kou JP, Liu SJ, et al. Qualitative and quantitative analysis of phenolic compounds in the leaves of *Aquilaria sinensis* using liquid chromatography-mass spectrometry. Phytochem Anal PCA. 2013;24(4):349–56. Epub 2013/03/14.

Zhang BP, Zhao J, Li SS, Yang LJ, Zeng LL, Chen Y, et al. Mangiferin activates Nrf2-antioxidant response element signaling without reducing the sensitivity to etoposide of human myeloid leukemia cells in vitro. Acta Pharmacol Sin. 2014;35(2):257–66. Epub 2014/01/01.

Zhang B, Zhao J, Li S, Zeng L, Chen Y, Fang J. Mangiferin activates the Nrf2-ARE pathway and reduces etoposide-induced DNA damage in human umbilical cord mononuclear blood cells. Pharm Biol. 2015;53(4):503–11. Epub 2014/11/7.

Zhao J, Zhang B, Li S, Zeng L, Chen Y, Fang J. Mangiferin increases Nrf2 protein stability by inhibiting its ubiquitination and degradation in human HL60 myeloid leukemia cells. Int J Mol Med. 2014;33(5):1348–54. Epub 2014/03/15.

Zhou M, Wang H, Suolangjiba, Kou J, Yu B. Antinociceptive and anti-inflammatory activities of *Aquilaria sinensis* (Lour.) Gilg. Leaves extract. J Ethnopharmacol. 2008;117(2):345–50. Epub 2008/03/21. http://www.ncbi.nlm.nih.gov/pubmed/?term=Suolangjiba[Author]&cauthor=true&cauthor_uid=18353573.

Zhu X, Cheng YQ, Du L, Li Y, Zhang F, Guo H, et al. Mangiferin attenuates renal fibrosis through down-regulation of osteopontin in diabetic rats. Phytother Res. 2015;29(2):295–302. Epub 2014/11/7.

Chapter 9
Acoustic-Based Technology for Agarwood Detection in *Aquilaria* Trees

Lina Karlinasari and Dodi Nandika

Abstract Agarwood is a non-timber forest product with high economic value. This incense wood is produced in the living stem of *Aquilaria* tree species after exposure to several stress factors and is not profoundly visible. Traditionally, agarwood hunters or collectors predict the existence of agarwood in the tree by relying on their visual assessment and experience, which may not be accurate at all times. To overcome this uncertainty, new technologies such as the nondestructive testing (NDT) technology are more reliable at providing valuable information for managing an agarwood plantation. In this study, an acoustic-based NDT has been applied on wounded and inoculated *Aquilaria* trees from which a reduction in sound velocity has been recorded. When coupled with PiCUS® sonic tomograph device, the tomogram results displayed progressing stages of decays consistent with that caused by fungal attack. The technology has been successful in detecting the existence of the agarwood in the target tree and can be improved for estimating the quantity and quality of agarwood in live trees periodically, before the final harvesting.

9.1 Introduction

Agarwood is a natural product, notably from the *Aquilaria* trees, which are largely native to Southeast Asia (Indonesia and Malaysia), east India and southern China. Its high value has led to overharvesting from the natural forests, resulting in *Aquilaria*, being listed as threatened species (IUCN Red List of Threatened Species, World Conservation Monitoring Centre 1998) and the trade of the product regulated since 2004 (Convention on International Trade in Endangered Species (CITES) of Wild Fauna and Flora 2004). Efforts have been made to conserve agarwood resources and to increase their supply through artificial cultivation. However, the pressure on natural *Aquilaria* trees is worrying because agarwood harvesting is

L. Karlinasari (✉) • D. Nandika
Department of Forest Products, Faculty of Forestry, Bogor Agricultural University (IPB), Bogor 16680, Indonesia
e-mail: karlinasari@ipb.ac.id

R. Mohamed (ed.), *Agarwood*, Tropical Forestry,
DOI 10.1007/978-981-10-0833-7_9

often done by trial and error, specifically, by cutting the tree without any reliable detection methods capable to confirm the presence of agarwood. Harvesting becomes excessive when the price of agarwood escalates, resulting in fatal harvesting through felling and chopping the whole tree. This practice contradicts sustainable forest management principles while at the same time is technically inefficient. One of the methods for collecting agarwood is direct harvesting of trees that are visually inspected by experienced agarwood collectors. However, indiscriminate tree felling in natural forests and among cultivated trees has been on the rise due to the impatience of greedy collectors. The healthy wood of *Aquilaria* tree is white, soft, and without scented resins. Generally, agarwood is formed in the living *Aquilaria* tree due to physiological changes that take place in the cells and secretion of chemical compounds into the wood when the tree is infected by fungi, wounded, or simply under stress. Agarwood is the resulting fragrant resin from the infected part of the tree, often deep inside the tree trunk. This chapter presents study concerning the utilization of nondestructive testing in field assessment based on acoustic technology. This technology has been widely used in tree quality assessment especially for detecting internal decay and defects in urban trees as well as assessing the quality of wood in trees (Brashaw et al. 2009).

9.2 Nondestructive Testing Technology for Wood-Decay Detection

Decay in wood is mostly caused by fungi and insect attack (Richardson 1993; Lebow 2010). The decaying agents will degrade woody tissues through the decomposition of cellulose and lignin. Numerous approaches have been attempted to detect internal decay condition within living trees in a minimally invasive manner. They are known as semi- or nondestructive testing/evaluation (NDT/E) methods. In many of the applications, probes, transducers, electrodes, etc. are inserted into the wood as a way to gain more information in investigating the wood decay. A number of techniques have been used to detect and assess decay in trees. Drilling techniques, such as in penetrometer, resistographs, increment coring (increment borer), and electrical sensors (e.g., Shigometer), involve drilling through the bark into the xylem (Costello and Peterson 1989; Smiley and Fraedrich 1992; Nicolotti et al. 2003; Wang and Allison 2008). Acoustic-based technique builds on the concept that stress wave (ultrasound or sound waves) propagation is sensitive to the presence of wood decay. Stress wave velocity is directly related to the physical and mechanical properties of wood. In general terms, stress waves travel slower in decayed or deteriorated wood when compared to sound wood (Pellerin and Ross 2002).

NDT technology based on sound propagation techniques has been successfully used to provide reliable information for evaluating trees (Yamamoto et al. 1998; Wang and Allison 2008; Lin et al. 2011; Indahsuary et al. 2014; Karlinasari et al. 2015a, b). The speed or velocity of sound as it passes through a tree is a common evaluation parameter (Pellerin and Ross 2002). The speed of sound can be affected by grain angles, knots, and advanced decay or deterioration, as well as microstructural characteristics, chemical composition, moisture content, and direction of wave propagation (i.e., longitudinal, radial, or tangential). For evaluating standing trees, the speed at which a wave travels across the diameter of the trunk (i.e., the radial measurement direction) is a good indicator of the presence of decay and other deterioration within the tree's trunk (Ross 1999; Sandoz et al. 2000).

The single-path stress wave technique was the first generation of stress wave-based equipment used in decay detection in wood (Pellerin and Ross 2002). The equipment used was a two-probe system that measures the wave transmission time in a single path. The capability of a single-path approach for tree decay detection has proven to be limited because stress wave velocity across tree stems varies substantially even for intact trees, and a standard reference velocity for data interpretation is not readily available (Wang et al. 2005).

Recently, a technology based on the measurement of acoustic wave velocity has been developed in tomographic investigation, whereby a cross section of an object is reconstructed to assess its internal condition (Nicolotti et al. 2003; Gilbert and Smiley 2004; Wang and Allison 2008; Deflorio et al. 2008; Wang et al. 2009; Brazee et al. 2011; Ahmad et al. 2012; Indahsuary et al. 2014; Li et al. 2014). The principle is based on the fact that sound passes more quickly through solid, sound, and high-quality wood than decayed, cracked, deteriorated, or low-quality wood. The instrument measures time and distance of sonic waves traveling through the wood and then the computer calculates the matrix of sound velocity and presents the velocity in the form of a tomography image, revealing locations and degrees of solid and damaged wood by different colors and percentages. Deflorio et al. (2008) detected a reduction in sound velocity when sycamore trees (*Platanus occidentalis*) were inoculated with decaying agents such as *Kretzschmaria deusta* and *Trametes versicolor*. Others have also reported of positive correlation between visual assessments and tomogram results. The latter has over 80 % accuracy at detecting wood decay compared to the former (Gilbert and Smiley 2004; Wang and Allison 2008; Wang et al. 2009; Brazee et al. 2011; Lin et al. 2011; Li et al. 2014).

9.3 Detecting the Presence of Agarwood

Early detection of internal decay in agarwood trees could provide a significant benefit to presume the presence of agarwood resins. Although visual assessment and experience of hunters or collectors can predict the existence of agarwood, sometimes external signs are not always obvious. For that, the NDT technology can be developed to assess the agarwood quality and estimates the volume of the resource.

9.3.1 Visual Detection

A tree that has been induced through fungal inoculation is often characterized by white areas on the trunk (mostly white spots), and black holes become evident (Fig. 9.1). During visual inspection at the locality of inoculation, agarwood is considered formed when the whitish healthy xylem turned into dark brown or is blackened (Fig. 9.2). In addition, the dark wood releases a pleasant aroma when burnt.

Fig. 9.1 *Aquilaria* trees that had been induced are marked with spots and holes (*red arrows*). (**a**) *Aquilaria microcarpa* plantation at Carita, Banten Province, (**b**) *Aquilaria malaccensis* plantation at Prabumulih, Palembang, South Sumatra, and (**c**) an *Aquilaria* tree planted at the BIOTROP garden, Bogor, West Java

Fig. 9.2 The discolored xylem (*brownish to black*) against the white fiber indicates agarwood presence. (**a**) *Aquilaria microcarpa* and (**b**) *Aquilaria malaccensis*

9.3.2 Development of Acoustic-Based Technology for Agarwood Detection

Commonly, collectors, hunters, as well as land managers are capable at identifying trees that contain agarwood based on experience. However, in the case one has to estimate the state of agarwood development and to make management decision as for the right time to harvest, one cannot depend solely on a human's perception. Data from supporting techniques are very valuable to ensure the resources do not go wasted. NDT techniques are seen as important tools that can give valuable information to the managers.

NDT methods have been used to detect and assess the progress of agarwood formation in *Aquilaria* trees. Here, we discuss studies that have been performed to evaluate the reliability of NDT, specifically using the acoustic tools. A study on a 2-year-old post-inoculated tree revealed that ultrasonic equipment based on time-of-flight technique is able to generate ultrasonic velocity that travels across two opposite sides of the trunk (Ahmad et al. 2012). The verification on felled tree confirmed that a thin resinous layer of agarwood was present between the decayed and solid parts of the trunk. The decayed part had velocity readings at low levels while in the sound part the sounds travels at high speed.

In a more comprehensive study, the acoustic technology was applied on 35 *A. microcarpa* trees (Indahsuary et al. 2014; Karlinasari et al. 2015a). Each tree had a diameter at breast height (dbh) in the range of 15–30 cm. The trees were planted in 1998 by the Forestry Research and Development Agency (FORDA), Ministry of Forestry, Republic of Indonesia, at Carita, Banten Province. In November 2009, the trees were inoculated with a strain of *Fusarium solani* (FORDA-CC 00509) following the fungal inoculation technology developed by FORDA (Chap. 4). Briefly, holes were drilled into the trunk at predetermined spacing from the base to the top. A certain amount of the fungus isolate was then inoculated into each hole.

Three years after the inoculation process, we applied the acoustic-based technology on the trees to evaluate the presence of agarwood. The single-path stress wave testing was conducted using the SylvatestDuo® acousto-ultrasonic, and the PiCUS® sonic tomograph recorded sonic wave velocity from several mounted sensors. The ultrasonic wave in SylvatestDuo® (frequency 22 kHz) as shown in Fig. 9.3 was generated by the device through a transmitting signal transducer, which travels across the trunk's diam-

Fig. 9.3 A field test using the NDT technology of Sylvatestduo® acousto-ultrasonic equipment

Fig. 9.4 A field test using the NDT technology of PiCUS® sonic tomograph

eter, and then the wave was received by a transducer. The sound velocity and time propagation in trees were revealed in the device screen. The study was conducted to measure at three heights (20, 130, and 200 cm above ground) and the ultrasonic wave velocities were recorded. The radial directions used were north–south and east–west.

Meanwhile, four to six sensors of the PiCUS® sonic tomograph (frequency between 1 and 3 kHz) as well as a number of nails were placed around the trunk in a horizontal plane. A sensor was magnetically attached to the nail. The sound was generated by tapping the nails three times with an electronic hammer, which was connected to the sensor and a laptop (Fig. 9.4). The nails serve as sending points and receivers. Sound velocities were captured and recorded (Fig. 9.5), and then a complete data matrix was processed to form a tomography image. The tomography image has a display of colors; dark brown suggests a high velocity, whereas low sonic velocities are denoted in blue color. Other colors of green to violet represent early and moderate fungus infection with various levels of rotting zones based on sonic velocity measurements in the respective areas.

A field assessment on 35 *A. microcarpa* tree species that had been inoculated showed the average value of the ultrasonic wave velocities (Vusn) was approximately 29.5 % higher than the sonic (Vsn) (Fig. 9.6) (Karlinasari et al. 2015a). The average values of radial Vsn and Vusn were 729 and 950 m s^{-1}, respectively. It also appeared that Vsn or Vusn values were not affected by a tree's height possibly because the induction treatment was applied in a uniform manner from the bottom to the top of the trunk including branches.

a

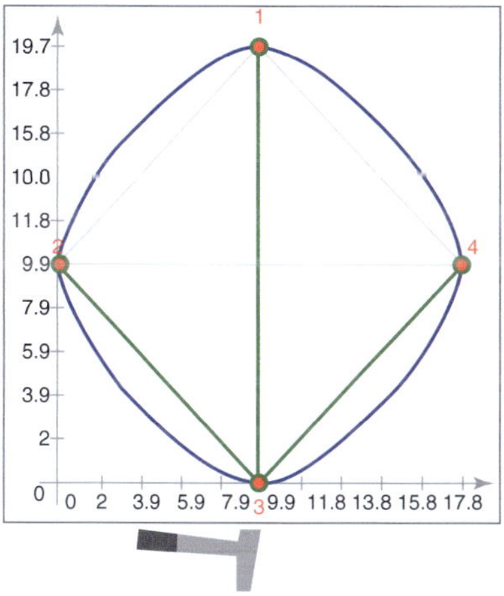

b

P/P	1	2	3	4
1	0	767	844	748
2	746	0	599	800
3	819	610	0	607
4	709	795	609	0

Fig. 9.5 PiCUS® sonic tomograph system: (**a**) tapping position on a sensor and the path taken by the wave and (**b**) data in the form of sonic velocities (m s^{-1})

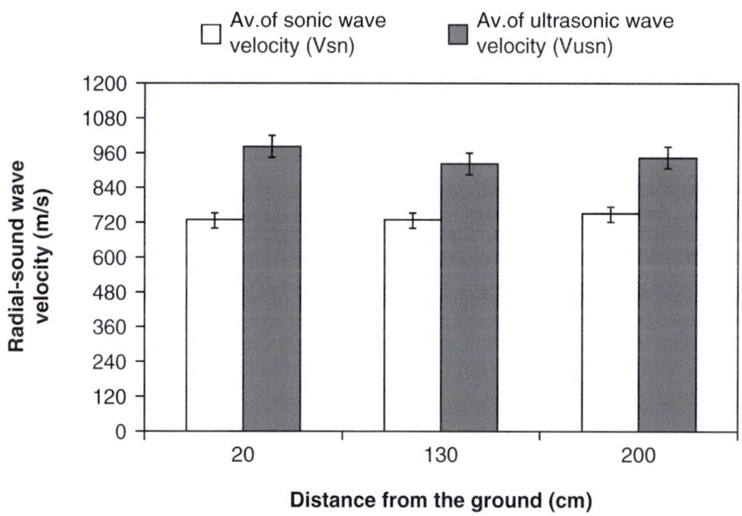

Fig. 9.6 The average sound wave velocities as recorded from the sonic and ultrasonic applications on 35 *Aquilaria microcarpa* trees (Source: Karlinasari et al. (2015a))

Three trees (designated as tree numbers 17, 19, and 32) were chosen based on natural signs, recommendation of experts, and tomogram results and felled to verify agarwood formation. Each trunk was then divided vertically into three parts following the radial measurement heights. The sound velocity average of those three trees was 692 m s^{-1} and 972 m s^{-1} for sonic and ultrasonic measurements, respectively

Table 9.1 Sound velocities were recorded at three different heights of *Aquilaria microcarpa* trees using sonic and ultrasonic wave testings

	Sound velocity (m s^{-1}) at three different heights from the ground			
Tree no.	20 cm	130 cm	200 cm	Average
Sonic wave testing				
17	725	733	779	746
19	649	706	702	686
32	638	668	629	645
Average				**692**
Ultrasonic wave testing				
17	940	990	1002	978
19	942	1029	1106	1026
32	945	849	943	912
Average				**972**

(Table 9.1). The moisture content and density of the fresh trunk were determined using an increment core sample (5 mm in diameter) at 5–10 cm around the dbh of each tree. The average values of green moisture content and density were 52.77 % and 0.76, respectively. Our results are in agreement with other research findings. For example, the average sonic velocity in a sound trunk of *Acacia mangium* is 1087 m s^{-1}, while that of a hollowed stem is 563 m s^{-1} (Yamamoto et al. 1998). Sandoz et al. (2000) reported that ultrasonic velocity between 600 and 1200 m/s corresponds to the decayed areas, while sound zones showed higher velocities, between 1200 and more than 2000 m/s. For some temperate tree species, the reference radial stress wave velocities in intact trees are in the range of 1100–1700 m s^{-1} (Divoz and Szalai 2002), whereas several tropical trees in forest plantations recorded the ultrasonic velocity at more than 1100 m s^{-1} (Karlinasari et al. 2012). A study by Oliveira et al. (2005) on Brazilian wood reported that the ultrasonic velocity in perpendicular to fiber direction increased about 7 % from green condition (MC 60 %) to dry condition (MC 6 %).

In another study, Karlinasari et al. (2015b) installed 6–12 PiCUS® sonic tomograph sensors on five inoculated *A. malaccensis* trees at different selected heights from the ground. The study was to evaluate internal conditions of those trees. The tree's cross sections revealed the condition of the tree trunk 30 months after inoculation with the *F. solani* inoculum of FORDA (CC00500) (Fig. 9.7a). A two-dimensional, tomogram images (Fig. 9.7b) display areas of different colors: dark brown, light brown, green, and violet, which indicate presence of various levels of rotting zones. Intact wood is dark brown while degraded zones are denoted by green to violet colors. It is obvious that the trunk was still intact but in some parts, the decaying process had started, most probably because of fungal attack. Figure 9.7c shown in 3D the distribution of internal condition along the height of inspection.

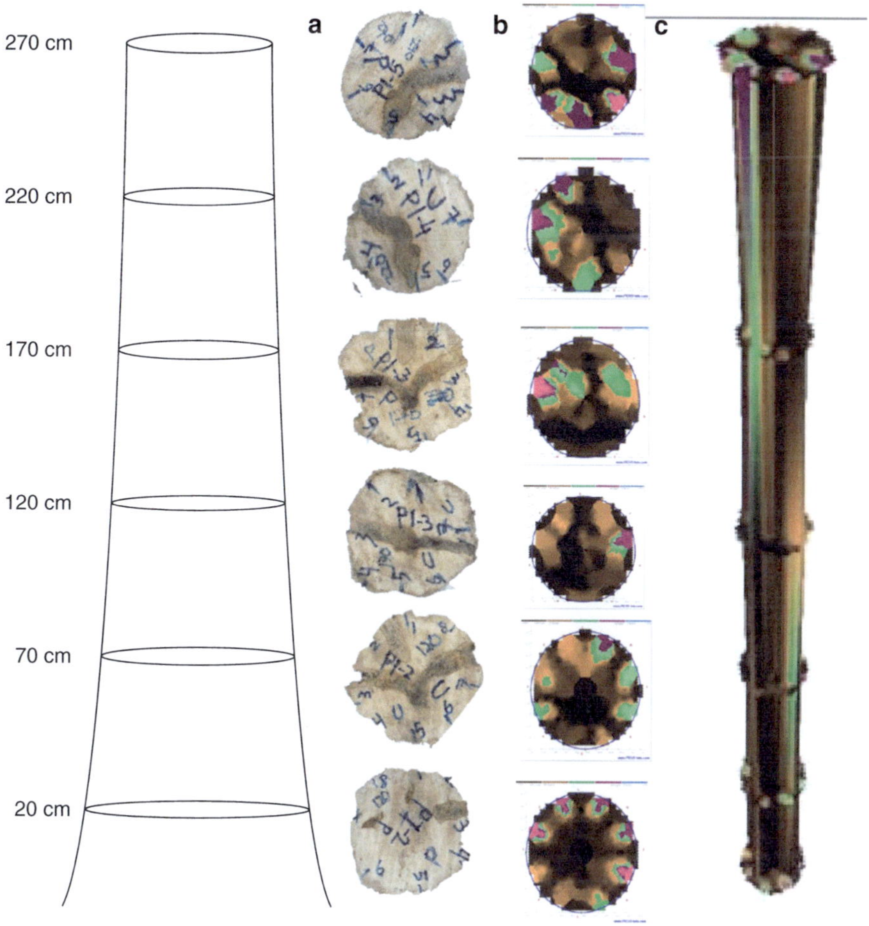

Fig. 9.7 Agarwood evaluation along the tree stem using the sonic tomograph at six heights aboveground. (**a**) Visual condition of the cross section at the various heights of tree, (**b**) tomogram images showing internal condition, (**c**) 3D of the tomogram images condition along the height of inspection

Nicolotti et al. (2003) reported that ultrasonic tomography technique was sensitive and effective for detecting internal decay at its early stage, locating with accuracy the position of the anomalies and estimating their size and shape. Meanwhile, the investigation on incipient decay using PiCUS® sonic tomograph for some temperate species (Douglas fir, beech, oak, and sycamore trees) inoculated with *Ganoderma applanatum*, *K. deusta*, and *T. versicolor* reported a decrease in sound velocity. However, detection efficacy of tomogram was host dependent (Deflorio et al. 2008).

Fig. 9.8 A trunk of *Aquilaria malaccensis* harvested 15 months after inoculation using FORDA's technique. (**a**) PiCUS® sonic tomograph sensors were installed on the trunk, (**b**) trunk diameter of 32 cm, and (**c**) visible presence of agarwood in the trunk

In a related study, we evaluated an *A. malaccensis* trunk, harvested 15 months after inoculation, belonging to FORDA collection (Fig. 9.8). The trunk was induced using FORDA's technique and agarwood production was verified from the aroma, physical appearance, and chemical testing. The tomogram results (Fig. 9.9a, b) demonstrated that the intact wood is still dominant, although fungal infection has indicated a forming of agarwood (green to violet). The average sonic velocities of that sample were 699 m s^{-1} and 729 m s^{-1} at heights of 30 cm and 85 cm, respectively, and the moisture content of this trunk was about 20 %.

9.4 Conclusion and Ongoing Research

The traditional way of detecting the presence of agarwood in *Aquilaria* sp. tree is by observing the natural signs such as white spots on the stem, dark or black marks in the xylem, as well as the aromatic smell of the resin. It is of common knowledge that black/dark wood color and strong aroma indicate the presence of agarwood.

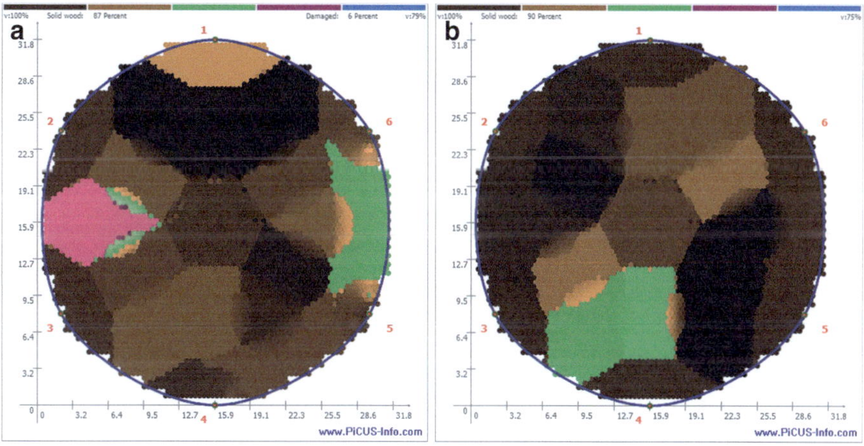

Fig. 9.9 Colored tomogram from the infected *Aquilaria malaccensis* trunk at heights of (**a**) 30 cm and (**b**) 85 cm aboveground

However, this practice is often destructive because some parts of the tree need to be harvested and it does not reflect the condition of the whole tree. In NDT approach, the instrument measures and provides important information regarding the internal health of the tree such as the interior decay events without having to damage the tree. This makes NDT technique, for example one based on acoustic technology attractive for use in verifying the existence of agarwood in standing trees. This study highlights the potential of the acoustic-based technology to detect the presence of agarwood in trees. In the near future, it is hoped that the research can be improved to estimate the quantity and quality of the resulting agarwood. During the degradation process, decaying fungi modify the chemical composition of the wood and may cause changes of ionic concentration and moisture content. This phenomenon influences resistivity and dielectric properties of wood. Further research is necessary to explore the electrical resistivity properties to determine other physical parameters in agarwood-producing trees.

References

Ahmad NI, Almuin N, Mohammad F. Ultrasonic characterization of standing tree. 18th World Conference on Nondestructive Testing. 2012; 16-20 April 2012. Durban, South Africa. http://www.ndt.net/article/wcndt2012/papers/515_Ahmad.pdf.

Brashaw BK, Bucur V, Divos F, Gonçalves R, Lu J, Meder R, Pellerin RF, Potter S, Ross RJ, Wang X, Yin Y. Nondestructive testing and evaluation of wood: a worldwide research update. Forest Prod J. 2009;59(3):7–14.

Brazee NR, Marra RE, Gocke L, Wassenaer PV. Non-destructive assessment of internal decay in three hardwood species of northeastern North America using sonic and electrical impedance tomography. Forestry. 2011;84(1):33–9.

CITES. Convention on International Trade in Endangered Species of Wild Fauna and Flora. 2004. 13th Meeting of the Conference of the Parties. 3–14 Oct 2004. Bangkok. Available online: http://www.cites.org/common/cop/13/raw_props/ID-Aquilaria-Gyrinops.pdf. Accessed on 15 Jan 2014.

Costello LR, Peterson JD. Decay detection in eucalyptus: an evaluation of two methods. J Arboriculture. 1989;15(8):185188.

Deflorio G, Fink S, Schwarze FWMR. Detection of incipient decay in tree stems with sonic tomography after wounding and fungal inoculation. Wood Sci Technol. 2008;42:117–32.

Divoz F, Szalai L. Tree evaluation by acoustic tomography. In: Proceedings of the 13th International Symposium on Nondestructive Testing of Wood. 19–21 Aug 2002. Berkeley; 2002. p. 251–6.

Gilbert EA, Smiley ET. PiCUS sonic tomography for the quantification of decay in white oak (*Quercus alba*) and hickory (*Carya* spp.). J Arboriculture. 2004;30(5):277–80.

Indahsuary N, Nandika D, Karlinasari L, Santoso E. Reliability of sonic tomography to detect agarwood in *Aquilaria microcarpa* Baill. J Indian Acad Wood Sci. 2014;11(1):65–71.

Karlinasari L, Maryanti IL, Batubara HN, Dhani RM, Nandika D. Evaluasi visual dan karakteristik kecepatan gelombang ultrasonik pohon peneduh di perkotaan dan hutan tanaman. Jurnal Ilmu dan Teknologi Hasil Hutan. 2012;5(2):40–6.

Karlinasari L, Indahsuary N, Kusumo HT, Santoso E, Turjaman M, Nandika D. Propagation of sonic and ultrasonic waves in agarwood trees (*Aquilaria microcarpa*) inoculated with *Fusarium* spp. J Trop Forest Sci. 2015a;27(3):351–6.

Karlinasari L, Nandika D, Turjaman M. Study on determination quality and quantity of agarwood using nondestructive testing technique (NDT) to support sustainable forestry management. In: Report – strategic national research grant from Ministry of Research, Technology, and Higher Education. Bogor Agricultural University (IPB); 2015b.

Lebow ST. Chapter 15. Wood handbook – wood as an engineering material. In: Wood Preservation. Madison: U.S. Department of Agriculture, Forest Service, Forest Products Laboratory; 2010.

Li G, Wang X, Feng H, Wiedenback J, Ross RJ. Analysis of wave velocity patterns in black cherry trees and its effect on internal decay detection. Comput Electron Agric. 2014;104:32–9.

Lin CJ, Cahnag TT, Juan MY, Lin TT. Detecting deterioration in royal palm (*Roystonea regia*) using ultrasonic tomographic and resistance microdrilling technique. J Trop Forest Sci. 2011;23(3):260–70.

Nicolotti G, Socco LV, Martinis R, Godio A, Sambuelli L. Application and comparison of three tomographic techniques for the detection of decay in trees. J Arboriculture. 2003;29:66–78.

Oliveira FGR, Candian M, Lucchette F, Salgon JL, Sales A. Moisture content effect on ultrasonic velocity in *Goupia glabra*. Mater Res. 2005;8(1):11–4.

Pellerin RF, Ross RJ. Nondestructive evaluation of wood. Madison: Forest Products Society; 2002.

Richardson BA. Wood preservation. 2nd ed. London: E & FN Spon; 1993.

Ross RJ. Using sound to evaluate standing timber. Int For Rev. 1999;1(1):43–4.

Sandoz JL, Benoit Y, Demay L. Standing tree quality assessments using acousto ultrasonic. In: Backhaus GF, Balder H, Idczak E, editors. International Symposium on Plant Health in Urban Horticulture, Braunschweig; 22–25 May 2000. p. 172–9.

Smiley ET, Fraedrich B. Determining strength loss from decay. J Arboriculture. 1992;18(4):201–4.

Wang X, Allison RB. Decay detection in red oak trees using a combination of visual inspection, acoustic testing, and resistance microdrilling. Arboriculture Urban Forestry 2008. 2008;34(1):1–4.

Wang X, Wiedenback J, Ross RJ, Forsman JW, Erickson JR, Pilon C, Brashaw BK. Nondestructive evaluation of incipient decay in hardwood logs. Gen. Tech. Rep. FPL-GTR-162. U.S. Department of Agriculture, Forest Service, Forest Products Laboratory, Madison; 2005. 11 p.

Wang X, Wiedenback J, Liang S. Acoustic tomography for decay detection in black cherry trees. Wood Fiber Sci. 2009;41(2):127–37.

World Conservation Monitoring World Centre. *Aquilaria microcarpa*. In: IUCN 2013. IUCN red list of threatened species. Version 2013.2. 1998. Available online: http://www.iucnredlist.org/details/33698/0. Accessed on 15 Jan 2014.

Yamamoto K, Sulaiman O, Hashim R. Nondestructive detection of heart rot of *Acacia mangium* trees in Malaysia. Forest Prod J. 1998;48(3):83–6.

Chapter 10
Keeping Up Appearances: Agarwood Grades and Quality

Rozi Mohamed and Shiou Yih Lee

Abstract Agarwood has many grades and goes by countless different names in both the sourcing and consuming countries. The different grades and classes of agarwood result from long-standing grading practices adopted by the people of each country. No standard method is available partly due to the intricacy during the hierarchical process of selling and buying. The foremost reason is the appearance of the traded agarwood itself, which can come in many forms from raw, such as chips, blocks, and flakes, to finished products such as oil, incenses, perfumes, accessories, and carvings. Agarwood in raw forms is of mixed quality; thus, the price and grade depend on this blended appearance. As the product is passed down from collectors to various levels of traders and finally to the buyers, the grade can be readjusted and the price inflated or understated depending on the interest. Therefore, buyers, traders, and collectors heavily rely upon time-honored trust when concluding a business deal. Authorities have not found the formula to standardize the grading system of agarwood trade, and this leads to the lack of coordination and regulation at international level. Nevertheless, several sourcing and consuming countries have made the effort to grade their agarwood according to their own local market, which can be used as a benchmark in formulating a more contemporary method that could be acceptable to all countries.

10.1 Introduction

Agarwood is a highly valuable commodity that has been traded in many parts of the world for hundreds of years. Overexploitation of the wild mother tree from which agarwood is derived affects regeneration of the young tree. As a result, *Aquilaria* and *Gyrinops*, the two major agarwood-producing genera, are currently

R. Mohamed (✉) • S.Y. Lee
Department of Forest Management, Faculty of Forestry, Universiti Putra Malaysia, UPM, Serdang, Selangor 43400, Malaysia
e-mail: rozimohd@upm.edu.my

R. Mohamed (ed.), *Agarwood*, Tropical Forestry,
DOI 10.1007/978-981-10-0833-7_10

listed under CITES protection (CITES 2014). Because agarwood (a bio-product from plant metabolism) is the outcome of a complex relationship between the tree host, infecting agent and environment, often the ensuing agarwood is of inconsistent quality. Even when induced in planted trees of uniform species and ages, the quality is difficult to predict. The sophisticated nature of agarwood is caused by many variable parameters; therefore, to determine its quality is a challenge. There appears to be a consensus among merchants, who habitually determine quality based on the country of origin, and the characteristics of the agarwood including the size, shape, color, scent and its durability, and the age and parts of the tree from where it is derived (Barden et al. 2000; Gunn et al. 2003; Wyn and Anak 2010). The Wild Trade Monitoring Network (TRAFFIC) reported that indicators for grading the quality of agarwood are really subjective and not proven (Antonopoulou et al. 2010). Many buyers also purchase their agarwood based on the country of origin and relates it to its quality. This might not be 100 % true, but their decision was established from experiences. For example, a consumer country like the United Arab Emirates (UAE) considers agarwood that originates from India and Cambodia to be of the highest quality followed by Malaysia, Laos-Myanmar, and Indonesia (Antonopoulou et al. 2010), while Japan regards Vietnam as the main source of *kyara*, the high-end agarwood grade, due to its supreme resin content (Compton and Ishihara 2004).

10.2 Agarwood Appearances

Agarwood is commonly traded in the form of chips. Other raw forms that are gaining popularity are blocks, logs, flakes, and powder, while the finished products in the international market are oil, incenses, perfumes, accessories, and carvings. The color and density of the wood are considered as important indicators of high-quality agarwood (Fig. 10.1a). Darker wood is believed to have higher amount of oleoresin compared to a less-darkened wood (Fig. 10.1b). The deep color also indicates that the agarwood comes from older trees. It has always been thought that agarwood from old growth contains the best quality. This appearance has been used by some sellers to entice buyers, by soaking the wood chips in a mixture of

Fig. 10.1 Different types of agarwood pieces and its imitation. (**a**) Grade A, (**b**) grade B, and (**c**) "black magic wood" (BMW)

petroleum-based synthetic oil so that it appears dark. This type of impregnated wood is known as the black magic wood (BMW) (Fig. 10.1c) (Antonopoulou et al. 2010; Wyn and Anak 2010; Gusmailina 2010; TRAFFIC East Asia-Taipei and TRAFFIC Southeast Asia 2005). Although initially considered a less-desired product, BMW now has a market of its own, especially in the UAE (Antonopoulou et al. 2010). For the ordinary people, BMW becomes an alternative to the expensive agarwood. BMW is reported to be traded from Indonesia (TRAFFIC Southeast Asia 2007).

Dark wood correlates with good-quality agarwood, but there is an exception. The highest grade of *kyara* has a light brown color and yet its quality is considered the greatest by Japanese consumers. Another quality indicator is the part of a tree from where the agarwood was removed. For example, an agarwood piece that came from the root is deemed more valuable when compared to other parts of the tree. Also, the resin amount in thick pieces is assumed higher than that of thin ones. It is difficult to ascertain agarwood quality by looking at its physical appearance. Some people have used the "sinking test," whereby wood chips are dropped in water to relate with the wood's quality. Wood that floats are considered of inferior quality due to the low resin content. However, this is easy to exploit by adding foreign material like metals into the wood for instance. A more popular method is the "burning test." Buyers can evaluate the aroma by burning a small sample of the agarwood chip. This method has its downside too as different people have different perceptions. Nevertheless, agarwood trading companies often hire trained personnel to sensory-evaluate the raw material before purchasing. Generally, many people agree that agarwood of the best quality has an outstanding aroma added with distinct odors when the wood piece is burned slowly or when the wood is distilled into essential oil. Some of the essential oils can be kept longer so that it develops the strong fragrance before being released into the market. The long-lasting fragrance indicates that the quality is superior, and naturally, it influences the grade and pricing (Antonopoulou et al. 2010; TRAFFIC East Asia-Taipei and TRAFFIC Southeast Asia 2005).

10.3 Traditional Knowledge

For many centuries, the indigenous people in Southeast Asia have depended on the forest for their livelihood. Forest was their source of food, income, and medicine. The indigenous people have inherited the knowledge about agarwood-producing trees from their ancestors and from the experiences they gained when exploring the forest. For the Penan Benalui indigenous people who live in the Borneo forest of Kalimantan, *Aquilaria malaccensis* and *Aquilaria microcarpa* are common agarwood producers. Trees that grow on the steep slopes or banks and along or near the streams are specifically sought-after as they contain agarwood of good quality (Donovan and Puri 2004). The indigenous people look for specific signs such as trees with shriveled leaf margins and brown leaf spots and having a "strange" smell. Crisp bark that came out from a slashed trunk is a good indicator of finding agarwood (Obidzonski 1997). At times, the indigenous collectors would gash open the trunk to check for agarwood. Only the part that contains agarwood will be excised. If there is no agarwood, often they will hack the trunk with a machete, an axe, or chisel, or peel the bark off, before leaving the tree undisturbed for several months to allow agarwood to form. Other signs include presence of ants, termites, or other nesting insects surrounding the wounded trunk; holes caused by insect's damage on the trunk or branches; evidence of a lightning strike; and knots on trunk or branches (Zich and Compton 2001; Gunn et al. 2003). One way to grade agarwood is based on the insect's pitch that is formed along the trunk; agarwood quality is said to correlate positively with the height of the pitch (Donovan and Puri 2004).

For some eager locals, they resort to chopping down the tree to inspect resin presence. They also dig out the root to expose the resin. This might be a quick way to look for agarwood; however, it eventually leads to low-quality agarwood production and death of the tree. Others build a makeshift ladder against the living tree to examine agarwood occurrence on upper trunk and branches. To collect agarwood, the dark wood will be extracted from the white wood using sharp knives, a broken glass, or hooked knives and dried under the shade (Zich and Compton 2001). Traditional collectors usually sell their agarwood to the middlemen, who in turn sell to the traders and retailers.

In consuming countries, most of the graders are well-experienced in determining the quality of agarwood. The experiences gained were mostly from traditional study and some inherited from their own ancestors. For example, in Japan, classification by incense masters is based on two criteria (Compton and Ishihara 2004). The first is the geographical source or place of origin, such as *kyara* (Vietnam), *rakoku* (Thailand), *manaban* (east-coast of India or Indo-Malaysia), *manaka* (Malaysia), *sasora* (western India), and *sumatora* (Sumatra, Indonesia). The second criteria is its taste such as sweet, sour, salty, hot, and bitter, which turns grading into a very subjective matter and people dependent.

10.4 Grading Systems in Sourcing Countries

Because there is no international common system for ranking agarwood, each sourcing country may have her own grading system or none at all. With the exception of China, most of the time, there is no government authority that actually sets up the rules and regulations pertaining to agarwood grades, let-alone standard. What is available are market grades that have been predetermined by sellers and buyers based on experiences. These grades are not necessarily accurate or inaccurate, but must be acceptable at both ends of the deal.

10.4.1 Malaysia

Agarwood grades in the Malaysian market heavily depended on physical appearances. There is no one grading system but generally, it is based on the ABC system. An example of this system is presented in Table 10.1 and is divided into nine grades (Mazlan and Dahlan 2010). Each grade may have additional subgrades, for example, grade A breaks down to A1 to A10. The prices are subjected to the perplexing grades and it can begin from anywhere between a few dollars to thousands of dollars per kg (Mokhtar et al. 2007; Wyn and Anak 2010). This huge range price can be inflated whenever there is increased demand and shrinking supplies.

In an attempt to better sort agarwood grades, Nor Azah et al. (2013) conducted chemical analysis on 34 agarwood samples conventionally categorized from super to low grades and recommended woodchip classification based on their resin content (Table 10.2). The resin content was calculated in % (w/w) and the process

Table 10.1 Commonly used agarwood grades found in the Malaysian market

Grade	Resin coverage on the surface	Resin color	Wood shape
Super king	Entire	Total black and shiny	Solid wood chunks (500 g to 3 kg)
Triple super	Entire	Total black and shiny	Solid wood chunks (200–500 g)
Double super	90 %	Less black and shiny	Solid wood chunks (50–200 g)
Super	80 %	Resin is black and grayish	Solid wood chunks of mixed sizes
A (A1-A10)	Entire	Black turning into gray	Solid wood chunks of mixed sizes
AB	Entire	Black turning into brown	Solid wood chunks of mixed sizes
B (B1-B10)	Entire	Black turning into brown	Solid wood chunks of mixed sizes
C	50 %	Gray	Varies in shapes and sizes
D	Entire	Gray and whitish	Varies in shapes and sizes

Source: Mazlan and Dahlan (2010)

Table 10.2 Agarwood classification based on resin content

Grade	Percentage of resin content
A	30 % and above
B	20–29.99 %
C	9–19.99 %
D	<9 %

Source: Nor Azah et al. (2013)

Table 10.3 Agarwood classification based on end products

Category	End use	Grade
Aroma	Wood chips and blocks containing fragrant resin for direct burning	Super, A, and B
Block	Wood blocks of various shapes and sizes, containing moderate to high density of fragrant resin, for use in making end products such as sculptures, beads, and bracelets	Tiger stripes and color of the sculpture
Classic	Wood blocks with fragrant resin of unique natural shapes for sale as an aesthetic product	Classic
Dust	Dust and debris, by-product of washing and oil extraction, but has remaining fragrant resin	Black, gray, yellow, dust, incense powder, debris
Extractable wood	Wood blocks and pieces of various sizes with low fragrant resin Suitable for oil distillation	C
Fragrance	Resin covers uniformly on one side of the wood pieces. Low to moderate fragrance	A1, A, and B

Source: MTIB (2014)

involves measuring constant weight of dried samples, dissolving the sample in ethanol, and then passing the filtrate through a reflux procedure followed by evaporation. Although the proposed method is quantifiable, the regulating body has not assimilated the system. Another disadvantage is that the method requires some extent of scientific analysis, which is not easily available to small-scale sellers and buyers.

The Malaysian Timber Industry Board (MTIB), as the governing body in Malaysia, is proposing a classification system (Table 10.3). By classifying agarwood products in relation to their end usage, it anticipates a standardized grade and an indicative price list of the product group will be brought to light.

10.4.2 China

Of all the sourcing countries, only China has its own dedicated system for grading agarwood, which started hundreds of years ago. In the old days, the submerging ability in water and aromatic odor appeared to be the main characteristics to grade agarwood quality. It is probably how agarwood derived its name, 沉香 (*chén xīang*), which means "sinking fragrance." The most comprehensive medical book in traditional Chinese medicine history, the Compendium of Materia Medica (*Bencao Gangmu*) written by Li Shizhen (1518–1593), revealed that agarwood has three "sinking" grades (Table 10.4). At present, some Chinese communities are still practicing the sinking test because it is simple and fast.

Nowadays, the grading and classification systems of agarwood in China have expanded to a greater length by including elements related to both arts and sciences.

Table 10.4 Agarwood classification in China using sinking ability

Grade	Name	Sinking level
1	水沉 (*shŭi chén*)	Total
2	栈香 (*zhàn xīang*)	Partial
3	黄熟香 (*húang shú xīang*)	Does not sink (floats)

Table 10.5 Modern classification of natural agarwood in China

Grade	White wood content	Resin coverage (%)
1	Not detectable	100
2	Little	>70
3	Relatively greater than Grade 2	>50
4	Large	>20

Table 10.6 Classification of natural agarwood in China based on elements of formation according to agarwood enthusiasts

Type	Category	Influencing elements	Fragrance degree
1	倒架 (*dăo jìa*)	Agarwood that was formed in broken branches or tree due to environment factors such as the storm	Light mellowness
2	土沉 (*tŭ chén*)	Agarwood that was formed in broken branches or tree that have been buried and decomposed by microbes	Heavy mellowness
3	水沉 (*shŭi chén*)	Agarwood that was formed in broken branches or tree that fell into a swampy area and have been biodegraded	Warm mellowness
4	蚁沉 (*yĭ chén*)	Agarwood that was formed in mechanically felled tree and left on the ground for termites consumption	Light reverberant
5	活沉 (*húo chén*)	Agarwood that was collected directly from a living tree	Elating
6	白木 (*bái mù*)	Agarwood that was collected from living tree of less than 10-years-old	Light scent

This is especially true for agarwood that is regarded as collectors' item. Based on common modern practices, natural agarwood from wild population falls under four grades (Table 10.5), whereby resin coverage on the wood is a distinct factor to classify the quality of the agarwood. Such practice is only common among experienced agarwood collectors and is not widely accepted as a standard to technically classify agarwood.

Besides resin coverage, another contributing factor to quality is the elements that concoct agarwood. Some believe agarwood that forms under different factors emits different types of scents (Table 10.6); however, each type does not reflect the quality itself. In fact, each one has its own exclusivity and whether it is worth for collection lies within the buyer's perspective. There could be more classes than what are listed here, as the classification is arbitrary and could be growing, as no official record is available.

In China, the main source of agarwood is from *A. sinensis*, which is also its endemic species. The plant species was listed in the China Plant Red Data Book in 1992 (Fu and Jin 1992) and further categorized in 1999 in the China's State Key Protected Wild Plants List (the first version) as Class II protected plant species. Agarwood is regarded as a valuable Chinese medicinal material. As a source of medicine, quality control of agarwood is often strictly monitored by the China Food and Drug Administration (CFDA 2004). In order to secure drug safety, measurements toward effective agarwood identification as medicinal material are recorded in the Chinese Pharmacopoeia.

In 1963, for the first time, the Chinese Pharmacopoeia (revised every five years) recorded agarwood as an essential Chinese medicinal material, under the name "Lignum Aquilariae" (Chinese Pharmacopoeia Commission 1963). Initially, the recognized agarwood sources were *Aquilaria agallocha* or *A. sinensis*, either from the wild or cultivated, with the former classified as imported agarwood from India, Malaysia, and other countries. The technique used to justify a suitable agarwood for medicinal use rely mainly on the color of the agarwood, the ability to submerge in water, and the presence of oil component surging out upon burning. However, the acceptance on imported agarwood was removed from 1995 onward, leaving only *A. sinensis* as the sole recognized source for Chinese medicinal applications (the Pharmacopoeia Commission of PRC 1995).

In 2010, the Chinese agarwood, also called "Aquilariae Lignum Resinatum," was formally described for use as Chinese medicines (CPC 2010). It has to be in the form of irregular pieces, flakes, or helmet-like shapes, with some in small-sized pieces. The surface should be uneven, with cut marks, sometimes with tiny holes and presence of visible black brown-colored resin stripes in between the pale yellow wood fiber. The tiny holes and dimple surface contain signs of decay. The quality of the product should be solid and sharp at edges and come with fragrant smell but bitter taste (CPC 2015). In addition, the CPC further regulated the quality by imposing four testing methods on the raw materials before qualifying them as suitable for Chinese medicines. These stringent requirements are imposed due to the high risk of using non-qualified agarwood for medicinal purposes. These methods are:

1. Wood anatomy

   A cross section of the wood should provide a clear view of 1–2 layers of ray parenchyma cells, fully containing brownish resin. The shapes of vessel outline and fiber outline are angular, with diameters between 42–128 μm and 20–45 μm, respectively. Cell walls are thin- to thick-walled in lignified cells. Included phloem is partially oblong or banded in shape, has thin cell walls, is non-lignified, contains brown resin, and is often in intercept with the ray parenchyma. It consists of loose fibers, and some fibers with thin cell wall may contain calcium oxalate crystals.

2. Alcohol-soluble extraction

   Based on a provided protocol, the final filtrate obtained from the extraction should contain oil of yellowish-brown color and emits fragrant scent. A drop of hydrochloric acid (HCl) is then added to the oil filtrate, together with a few drops of vanillin and absolute ethanol, rendering the extract into peach red color.

3. Thin-layer chromatography (TLC)

   TLC is the simplest and fastest method to identify the chemical content of a mixture by using the concept of capillary action. A small amount of sample is applied at the bottom of the solid adsorbent-coated TLC plate, placed in a shallow pool of solvent, and let to rise up via capillary action. When the solvent reaches equilibrium, the fluorescent dyed TLC plate will be visualized under the UV light at 365 nm. Using the TLC method, the color of the fluorescent spot of an unknown agarwood is compared to that of an authentic Chinese agarwood.
4. High-performance liquid chromatography (HPLC)

   In the updated version of CPC (2015), HPLC technique has been included as an analytical tool to separate, identify, and quantify each component available in a mixture. Compared to TLC, this technique provides an even more precise and sensitive identification level. There are two different ways to verify agarwood samples using the HPLC method:

   (i) Fingerprinting chromatography

      Based on the provided protocol on sample preparation, a standard chromatogram program setting is used to identify agarwood sample. Genuine agarwood sample should display six peaks, with three specific to agarwood: peak-1 is agarotetrol, peak-3 is 8-chloro-2-(2-phenylethyl)-5,6,7-trihydroxy-5,6,7,8 chromone, and peak-5 is 6,4′-dihydroxy-3′-methoxy-2-(2-phenylethyl) chromone. Peak-1 should have the same peak retention time as the standard, which is between the 19th and 20th minute.

   (ii) Content quantification

      Based on the provided protocol on sample preparation, a standard chromatogram program setting is used to quantify the content volume for agarotetrol ($C_{17}H_{18}O_6$). This method utilized liquid agarotetrol with ethanol as reference standard for quantification. The expected volume of agarotetrol content obtained from the dried sample should not be less than 0.10 %.

Although the recognition for imported agarwood was removed from the *Chinese Pharmacopoeia* in 1995, the CFDA issued a revised standard on the imported agarwood, namely, *A. agallocha*, in 2004. The qualification technique is the same as those stated in the *Chinese Pharmacopoeia*, but with stricter criteria. The imported wood permitted should not contain any coloring dye, wax, and non-resinous woody parts. The moisture content should not be above 10.0 % and the alcohol-soluble extraction using the hot bath method should provide extract of more than 15.0 % of its product (CFDA 2004).

Besides its use in medicine, modern Chinese people like to collect agarwood as a hobby. They view agarwood as a valuable ornamental display product, a product that is both unique and expensive and therefore only accessible to people of high-class society. They grade agarwood based on their personal assessment and experiences. The characteristics of a sought-after agarwood collection are often related to the beauty of the product, in which four elements are taken into account. A desirable agarwood should emit fragrance of clear, refreshing, mellow, and long-lasting characteristics. Besides, the oil content should be of high level. The shape, surface texture, grain, and

Table 10.7 Description of agarwood types in Indonesia

Types	Description
Gubal	Wood that contains agarwood with high amount of resin, black or dark brown in color, and has a strong aroma
Kemedang	Aromatic wood contains agarwood with less resin, gray or brown in color, and has rough fibers within soft wood
Agarwood powder	Powder that processed from agarwood chips or pieces

Source: Subehan et al. (2005), SNI (2011)

resin distribution are all important factors. There is no significant justification upon this matter as it solely relies on personal preferences. The agarwood piece can be natural or a sculpture. Finally, collectors view the color of the piece with great importance. Many consider not only the color of the resin but also the blending between the resin and the white parts as an artistic impression.

10.4.3 Indonesia

Indonesia is an important agarwood producer and it adopts similar approaches as other Southeast Asian countries in grading practices. The Indonesian authority, the National Indonesian Standard (SNI), proposed a grading system based on physical appearances such as the color, size, contamination of wood, density, and burned aroma (Subehan et al. 2005; SNI 2011). There are three types of agarwood in the Indonesian market: *gubal*, *kemedang*, and agarwood powder (Table 10.7), and each type has its own grades or quality (Table 10.8). A standard procedure for sampling and conducting investigations for grades determination was also prepared by SNI.

10.4.4 Vietnam

Agarwood was part of the tribute paid by Vietnam to the imperial court in Beijing in the early 19th century (TRAFFIC East Asia-Taipei and TRAFFIC Southeast Asia 2005; Chung and Purwaningsih 2008). No doubt that agarwood was treated as a select tax and a tribute item similar to gold (Jung 2013). However, there is no official record on agarwood grading system in Vietnam at present. What is currently being practiced by agarwood researchers, collectors, and traders is actually mentioned in the ancient books from Vietnam (Table 10.9).

Although the unofficial agarwood grading system is well-informed among agarwood industry players, yet it is not thoroughly applied by all parties. There are local agarwood companies that carry out their internal grading system, for example, grades special (best), followed by A, B, and C. There are also unofficial agarwood grading practices from outside Vietnam that grade Vietnam's agarwood from their origins and color of soil (Table 10.10).

Table 10.8 Commonly found agarwood grades in the Indonesian market

Type	Grade	Color	Weight	Aroma (burned)
Gubal	Double super	Equally black and shiny	Sink	Soft aroma (*wangi halus*)
	Super A	Unequal black and shiny	Sink	Soft aroma (*wangi lembut*)
	Super B	Black and not shiny	*Melayang* (float)	Aromatic
	Super middle A (under water)	Black	*Melayang* (float)	Aromatic
	Super middle A (up water)	Black	Float	Aromatic
Kemedangan	*Sabah*	Brownish black	*Melayang* (float)	Aromatic
	Kemedang A	Brown with black line	*Melayang* (float)	Aromatic
	Tanggung C	Brown with white narrow line	Float	Aromatic
	Kemedangan hijau	Brown with green line	*Melayang* (float)	Aromatic
	Kemedangan putih	Gray with black narrow line	Float	Pungent aroma (*wangi pedas*)
Agarwood powder	*Gubal* powder	Black	–	Aromatic
	Kemedangan powder	Whitish brown	–	Quite aromatic

Source: SNI (2011)

The evaluation criteria are strongly subjective because they are based on having knowledge about the origin of the agarwood and assessing its appearance such as color, taste, and oil content. In addition, experience, innate feeling, and agreement between both parties are the main keys to establishing agarwood grades in Vietnam (Ông and Nguyễn 2014). Although the trees' growing environment may have influence on the quality of the agarwood produced, yet place of origin and color of soil cannot precisely indicate the grades (Nguyễn V.M., pers. comm. 2015)

10.4.5 Papua New Guinea

The agarwood of Papua New Guinea is commercially graded into super A, A, B, C, D, and E classes according to physical appearances of the wood (Table 10.11) (Gunn et al. 2003). There are instances where the price of agarwood is not constrained by the grading system. Grade C of mixed chips has reportedly fetched a higher price than the large blocks of grade A or B. This could be explained by the market demand for woodchips.

Table 10.9 Agarwood grading list from ancient books in Vietnam

Class	Name	Description	Grade	Type/product
1st	*Kỳ nam*	Oil coverage is at maximum coverage, contains a number of flavors such as sour, spicy, bitter, and sweet. It emits special smell and green smoke which flows straight and long into the air when burnt	1	*Bạch kỳ* (white *Kỳ nam*)
			2	*Thanh kỳ* (green *Kỳ nam*)
			3	*Huỳnh kỳ* (yellow *Kỳ nam*)
			4	*Hắc kỳ* (Black *Kỳ nam*)
2nd	*Trầm*	Oil coverage is lesser compared to first class. Bitter taste. The fragrant is only smelled by burning, and it emits white smoke which flows nondirectionally and then vanishes	1	*Hoàng lạp trầm* (dark yellow agarwood)
			2	*Hoàng trầm* (yellow agarwood)
			3	*Giác trầm* (agarwood formed on wound surface)
			4	*Tiến hương* (sweet agarwood)
			5	Kê cốt hương (red agarwood)
3rd	*Tốc*	Little oil coverage, mostly obtained from the outer layer of the stem and cutting along the grain	n/a *Tốc đĩa is known to be of higher-quality level compared to others. However grading does not really applied in Tốc*	*Tốc đĩa* (disc *Tốc*, hairline resin formed on a 1-year wound left from previous harvesting)
				Tốc dây (wire *Tốc*, agarwood collected from *Aquilaria* trees growing at sea coast area. The trees do not grow straight and contains agarwood at swollen stems)
				Tốc hương (incense *Tốc*, agarwood used as fragrance incense)
				Tốc pi (*Pi Tốc*, agarwood that is in the form of tiny black spots)

Source: Ông and Nguyễn (2014)

10.5 Quality of the Essential Oil

Agarwood is traded largely in its natural form, which is woodchips. Woodchips of low grades and dust are generally processed into essential oil. Usually, distillation is the common method used by the industries and local people to produce agarwood

Table 10.10 Unofficial grading list of agarwood from Vietnam by Chinese agarwood collectors

Grade	Province	Region	Soil color	Agarwood color
Kỳ nam	Khánh Hòa Lâm Đồng Ninh Thuận	South Central Coast	Red	White Green Yellow Black
Super	Khánh Hòa Lâm Đồng Ninh Thuận	South Central Coast	Red	Dark yellow, yellow Red
Best	Quảng Nam (Phước Sơn, Hội An)	South Central Coast	Red	Dark yellow
Standard 1	Quảng Bình Thừa Thiên–Huế (Huế)	North Central Coast	Black	Red
Standard 2	Nghệ An Hà Tĩnh	North Central Coast	Black	Black
Lower	Gia Lai Kon Tum Đắk Lắk	Central Highlands	Yellow	Brown

Source: Anon (2014)

Table 10.11 Commonly used agarwood grades found in Papua New Guinea market

	Form/shape			
	Heavy irregular	Heavy regular	Light large pieces	Heavy thick chips
Color	Grades			
Black shiny	Super A+	A	B	C
Mixture of dark black and chocolate brown	B	B	C	C
Mixed color (pale black/ chocolate brown)	C	C	C	C
Brown	D	D	D	D
Pale yellow or tan brown	D mostly rejected	D mostly rejected	D mostly rejected	D mostly rejected
White	Reject	Reject	Reject	Reject

Source: Gunn et al. (2003)

oil (Persoon 2007; Jaapar 2008; Chetpattananondh 2012). The essential oil is mostly used in making perfumes and cosmetic products. In the Arab culture, for example, agarwood is highly regarded because there is a saying that the Prophet Muhammad had used it to incense his clothes. Therefore, agarwood fragrance has been associated to spiritual cleansing and applying agarwood- or *oudh*-based perfume became part of the Muslim culture (Antonopoulou et al. 2010). Essential oil from agarwood has been used to produce novelty fragrance by mixing it with oils from other fragrance-producing plants such as sandalwood, cinnamon, and cloves, to create unique pleasing scents that are now entering many market places in Asia

Table 10.12 Classification of agarwood essential oil

Grade	Description
A+	100 % oil purity
A	95–99 % oil purity
B	<95 % oil purity

Source: Chetpattananondh (2012)

and is catching non-Asian societies like Europe (Barden et al. 2000; Compton and Ishihara 2004).

The essential oil has many grades. How one oil grade differs from the other often depends on sensory evaluation and individual preferences. Similar to the woodchips, there appears to be general convention on the quality of agarwood oil. For example, agarwood traders from the UAE process the oil into three types of *oudhs*: (1) the expensive pure *dihn al oudh*, (2) the relatively cheaper *dihn al oudh*, and (3) the low-quality *oudh* like the blended Arabian perfumes such as *attars*, *mukhalats*, and *bakhoor* (Antonopoulou et al. 2010). Due to a lack of common grading standard or method for agarwood oil, Chetpattananondh (2012) classified agarwood oil into three grades based on oil purity (Table 10.12).

Several have reported on scientific approaches to overcome conventional agarwood oil classification, which in a way is parallel to woodchips – physical appearances, including oil color, density, and scent. The fact is that agarwood has different grades because of its compounds; high-quality agarwood contains more compounds compared to lower quality (Ismail et al. 2014; Jayachandran et al. 2014; Jong et al. 2014). For scientists, it is possible to analyze specific chemical compounds in the agarwood oil using sophisticated detection system such as gas chromatography-mass spectrometry (GC-MS) and HPLC. In fact, most international companies have their in-house detection system for analyzing compounds in the agarwood samples. High-quality agarwood contains two important compound types known as sesquiterpene and 2-(2-phenylethyl)-chromone (Naef 2011). GC-MS analysis of agarwood compounds is discussed in detail in Chap. 7.

However, to apply GC-MS/HPLC techniques is complicated and requires sample preparation. Hidayat et al. (2010) proposed the electronic nose approach using a sophisticated detection system to distinguish different grades of agarwood oil. The sample was first diluted with a solvent and placed in a stoppered vial, which was connected by inert tubing to the equipment. The vial was heated in a heater block and the volatiles were absorbed and detected by the electric nose sensor. This approach requires the implementation of statistical analysis such as the hierarchical cluster analysis (HCA) and principal component analysis (PCA) and must be tested on many different samples before it can be used. Once established, the artificial neural network can predict the identity of unknown agarwood samples including the region of origin (Hidayat et al. 2010; Najib et al. 2012; Ismail et al. 2013). Another technique that applies sophisticated instrument is the X-ray micro-computed tomography (micro-ct). The instrument is used to detect oleoresin distribution in the wood. The X-ray interacts with the solid matter and displays the wood structure from the white and black images (Van Den Bulcke et al. 2009; Yazid et al. 2009). This technique of radiation imaging is one of the alternatives to visualize the oleoresin in the wood chips without disruption.

The distillation method is both time- and energy-consuming; therefore it is not cost-effective. There are initiatives to develop new methods to produce agarwood oil that is high in both yield and quality. Such method uses the supercritical fluid (SF) CO_2 extraction, which has been proven to yield higher amount of oil compared to only 50 % from traditional extraction technique (Sepahpour et al. 2013). Thailand and Vietnam are two countries that have conducted advance research in agarwood oil production using the CO_2 technique. This technique is capable of producing high-quality oil with minimum waste residues but the cost to develop a factory with such technology could be an issue.

10.6 Agarwood Trade

India was the first country and the main supplier of agarwood for the international market until the end of 1990s. Later, in the mid-1990s, Cambodian agarwood became popular, but then, due to the political situation and smuggling activities, agarwood supplies from Cambodia became depleted. Due to demands, the locals passed mixed agarwood originated from other countries as Cambodian. Overexploitation of the trees has reduced their numbers in the wild habitat in affected countries such as India, Bangladesh, Cambodia, China, Thailand, and Vietnam. Nowadays, attention has shifted to Indonesia and Malaysia for fulfilling wild agarwood demands. Another potential source of wild agarwood is Papua New Guinea. Due to this long-standing threat, it is crucial to manage the agarwood-producing species in these countries in a sustainable manner. India, China, Vietnam, and Thailand now have successful agarwood tree plantations. It would be no surprise if one day these plantation species become major suppliers of agarwood replacing the wild resources. Demand for agarwood comes from many countries in the world. The Gulf Cooperation Council (GCC) includes countries such as Saudi Arabia, UAE, Kuwait, Oman, Bahrain, and Qatar, and is the dominant market for agarwood. Besides consumers in GCC countries, another group of end users comes from Hong Kong, Japan, Republic of Korea, and Taiwan. Their preferred agarwood products are in the form of incense, sculpture, natural shapes, bracelets, beads, and many more. These items are treasured because they are used in their cultural and religious rituals. While in China, the main demand of wild agarwood is for medicine and personal artifacts.

Several countries are involved in oil production like the UAE, India, Laos, Vietnam, Japan, Thailand, and Malaysia. Usually, businesses from these countries have their own oil factories and their source of raw agarwood comes from either plantations or other growers and intermediaries. Manufacturers from Japan and UAE prefer to select, purchase, and process the raw materials themselves and prevent the mixing of both high-grade and lower-grade qualities of agarwood chips. Singapore and UAE are known as trading centers or hubs for agarwood products,

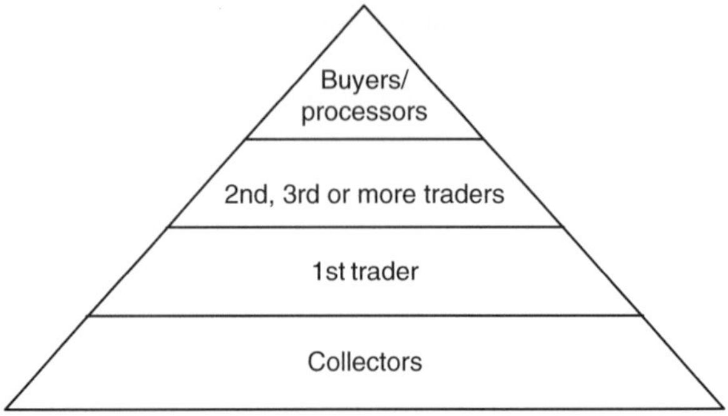

Fig. 10.2 Chain of custody in agarwood trade

while Dubai and Hong Kong are known as major market hubs. Singapore is importing and reexporting mostly from Malaysia and Indonesia to UAE and other countries. UAE plays a role as an entrepot and reexporting for the GCC and European countries. Since Dubai and Hong Kong are popular among tourists, businesses often seek opportunities to sell their agarwood products there.

There is a pyramid structure in agarwood trading starting with the locals as collectors at the very bottom of the pyramid. They represent the highest number of participants, who sell the raw product to middlemen (Fig. 10.2). The first trader buys different sizes of similar quality agarwood in small quantities from the collectors and sells it to the second trader. Several stages of intermediaries may occur, and as the chain continues, the quantity grows, until the agarwood reaches the final buyer. The buyers of the raw materials, at the top of the pyramid, constitute the least number of participants in the chain. They usually have their own factories to process the raw agarwood into end products that are tailored to meet demands in each consuming country.

10.7 Conclusion and Future Perspectives

In conclusion, agarwood grading is very subjective and relies on visualization and personal experiences. A standard grading system is absent across the sourcing region; nevertheless; each country has a customary market grading system in place. To substitute this system with one that is quantifiable would not be tempting to the players at the bottom of the chain (collectors) due to inaccessibility to modern techniques and the cost related to such system. Agarwood business itself is unique as it has begun centuries ago and has been very dependent on the mutual understanding between the buyers and the sellers. Because of the purchasing power, the grading system for agarwood is likely to fall under the prerogative of the consumer countries. Different consumer countries uphold different agarwood traditions. Japanese who practices agarwood appreciation has their own opinions about what a high-quality agarwood is while Chinese make sure that the raw agarwood meets the requirements set in the Chinese Pharmacopoeia before it can be applied in Chinese medicines. Among all the sourcing countries, only China has a strong domestic agarwood trade due to the fervent agarwood culture practices, while others are solely producing countries, which rely mostly on international trades. Due to the long history in international agarwood trade, the consumers from the importing countries seem to have developed preference in their agarwood species: Japanese, *A. crassna*; Chinese, *A. sinensis* and *A. crassna*; and Arabs, *A. malaccensis*. A grading system appears to be more applicable in consumer countries because they determine what suit them best. On the other hand, the sourcing countries may develop their own grading system that can contribute in standardizing their internal agarwood trade circulation and tailor it to the importing countries' requirements. By having a comprehensive method of grading to determine their agarwood quality, sellers will have better bargaining power. Therefore, it is useful for each sourcing country to develop a more systematic grading system aimed at meeting demands of consumers in targeted countries.

References

Anon. 全世界沉香品种分类大全 2014 (In Chinese).

Antonopoulou M, Compton J, Perry LS, Al-Mubarak R. The trade and use of agarwood (oudh) in the United Arab Emirates. TRAFFIC Southeast Asia; 2010.

Barden A, Awang Anak N, Mulliken T, Song M. Heart of the matter: agarwood use and trade and CITES implementation for *Aquilaria malaccensis*. Cambridge: TRAFFIC International Publication; 2000.

CFDA (China Food and Drug Administration). Quality standards on catechu and 43 other kinds of imported medicinal materials. Beijing: People's Health Press; 2004.

Chetpattananondh P. Overview of the agarwood oil industry. IFEAT International Conference in Singapore; 2012. p. 131–8.

Chung RCK, Purwaningsih. *Aquilaria malaccensis* Lamk. In: Oyen LPA, Nguyễn XD, editors. Plant Resources of South-East Asia 19: Essential-oil Plants. Jakarta: LIPI Press; 2008. p. 64–7.

CITES. Checklist of CITES species. 2014. http://checklist.cites.org/#/en/search/output_layout=alphabetical&level_of_listing=0&show_synonyms=1&show_author=0&show_english=1&show_spanish=1&show_french=1&scientific_name=aquilaria&page=1&per_page=20. 2 Dec 2014.

Compton J, Ishihara A. The use and trade of agarwood in Japan. Cambridge: TRAFFIC International; 2004.

CPC (Chinese Pharmacopoeia Commission). Pharmacopoeia of People's Republic of China, vol. 1. Beijing: People's Health Press; 1963.

CPC (Chinese Pharmacopoeia Commission). Pharmacopeia of the People's Republic of China. 1. Beijing: China Medical Science Press; 2010.

CPC (Chinese Pharmacopoeia Commission). Pharmacopeia of the People's Republic of China. 1. Beijing: China Medical Science Press; 2015.

Donovan DG, Puri RK. Learning from traditional knowledge of non-timber forest products: Penan Benalui and the autecology of *Aquilaria* in Indonesian Borneo. Ecol Soc. 2004;9(3):3. [online] URL: http://www.ecologyandsociety.org/vol9/iss3/art3.

Fu LK, Jin JM. China plant red data book-rare and endangered plants, vol. 1. Beijing: Science Press; 1992.

Gunn B, Stevens P, Singadan M, Sunari L, Chatterton P. Resource management in Asia-Pacific: eaglewood in Papua New Guinea. First International Agarwood Conference; 2003. p. 1–18.

Gusmailina G. Peningkatan Mutu Pada Gaharu Kualitas Rendah. Jurnal Penelitian Hasil Hutan. 2010;28(3):291–303.

Hidayat W, Shakaff AYM, Ahmad MN, Adom AH. Classification of agarwood oil using an electronic nose. Sensor. 2010;10:4676–85. http://creativecommons.org/licenses/by/3.0/ [18 Nov 2014].

Ismail N, Ali NAM, Jamil M, Rahiman MHF, Tajuddin SN, Taib MN. Differentiating agarwood oil quality using artificial neural network. Malays J Anal Sci. 2013;17(3):490–8.

Ismail N, Ali NAM, Jamil M, Tajuddin MHFRN, Taib MN. A review study of agarwood oils and its quality analysis. Jurnal Teknologi (Sci Eng). 2014;68(1):37–42.

Jaapar AJ. Extraction of gaharu essential oil using ultrasonic assisted steam distillation. Universiti Malaysia Pahang: Unpublished degree dissertation; 2008.

Jayachandran K, Sekar I, Parthiban KT, Amirtham D, Suresh KK. Analysis of different grades of agarwood (*Aquilaria malaccensis* Lamk.) oil through GC-MS. Indian J Nat Prod Resour. 2014;5(1):44–7.

Jong PL, Tsan P, Mohamed R. GCMS analysis of agarwood extracts from healthy and infected woods of mature and juvenile *Aquilaria malaccensis*. Int J Agric Biol. 2014;16(3):644–8.

Jung D. The cultural biography of agarwood – perfumery in Eastern Asia and the Asian neighbourhood. J R Asiatic Soc. 2013;23(1):103–25.

Malaysian Timber Industry Board (MTIB). Panduan pengelasan produk gaharu Malaysia. 1st ed. Kuala Lumpur: Lembaga Perindustrian Kayu Malaysia; 2014.

Mazlan M, Dahlan T. Pengredan dan pemprosesan gaharu. Selangor: Seminar Kebangsaan dan Pameran Gaharu; 2010.

Mokhtar Z, Tahir R, Aziz AA, Sinin S, Mahmud MKA. Gaharu: masa hadapan perladangan pokok karas di Semenanjung Malaysia. Kuala Lumpur: Persidangan Gaharu Kebangsaan; 2007.

Naef R. The volatile and semi-volatile constituents of agarwood, the infected heartwood of *Aquilaria* species: a review. Flavour Fragr J. 2011;26:73–89.

Najib MS, Ali NAM, Arip MNM, Jalil MA, Taib MN. Classification of agarwood using Ann. Int J Electric Electron Syst Res. 2012;5:20–34.

Nor Azah MA, Saidatul Husni S, Mailina J, Sahrim L, Abdul Majid J, Mohd Faridz Z. Classification of agarwood (gaharu) by resin content. J Trop Forest Sci. 2013;25(2):213–9.

Obidzonski K. Can gaharu be harvested sustainably? A case study from East Kalimantan, Indonesia. Georgia J Ecol Anthropol. 1997;1:28–35.

Ông HC, Nguyễn VM. Agarwood: benefits, challenges and prospect. In: Vietnam Agarwood Association (VAA), editor. Summary of agarwood planting techniques and developments. Vietnam Union of Science and Technology; 2014. p. 9–25.

Persoon GA. Agarwood: the life of a wounded tree. IIAS Newslett. 2007;45:24–5.

Sepahpour S, Selamat J, Manap Y. Extraction of agarwood essential oil using supercritical fluid extraction (SF-CO_2). Selangor: 1st International Scientific Symposium on Agarwood (ISSA): Agarwood in the new era; 2013. p. 33.

Standardisasi Nasional Indonesia (SNI). Gaharu SNI 7631: 2011 (SNI. 01–5009.1–1999). Indonesia: Badan Standardisasi Nasional; 2011.

Subehan JU, Fujino H, Attamimi F, Kadota S. A field survey of agarwood in Indonesia. J Trad Med. 2005;22:244–51.

The Pharmacopoeia Commission of PRC. Pharmacopoeia of the People's Republic of China, 1995's edition, vol. 1. Guangzhou: Guangdong Science and Technology Press, Beijing: Chemical Industries Press; 1995.

TRAFFIC East Asia-Taipei and TRAFFIC Southeast Asia. The trade and use of agarwood in Taiwan, Province of China. CITES; 2005.

TRAFFIC Southeast Asia, editors. Capacity-building workshop for improving implementation and enforcement of the cites listing of *Aquilaria malaccensis* and other agarwood-producing species. Proceedings of the experts group meeting on agarwood. Organized by TRAFFIC Southeast Asia. Kuala Lumpur, 14–17 Nov 2006.

Van Den Bulcke J, Boone M, Van-Acker J, Stevens M, Van-Hoorebeke L. X-ray tomography as a tool for detailed anatomical analysis. Ann Forest Sci. 2009;66(5):1–12.

Wyn LT, Anak NA. Wood for trees: a review of the agarwood (gaharu) trade. Kuala Lumpur: TRAFFIC Southeast Asia; 2010.

Yazid KI, Abdullah MZ, Saleh JM, Awang MR. Detection and distribution analysis of oleoresin in gaharu wood by X-ray micro-computed tomography (Micro-Ct). Kuala Lumpur: Persidangan Kebangsaan Penilaian Ekonomi Sumber Hutan; 2009. p. 250–7.

Zich F, Compton J. Towards sustainable management of Papua New Guinea's agarwood resource. New South Wales: A TRAFFIC Oceania report in conjunction with WWF South Pacific Programme; 2001.

MIX
Papier aus verantwortungsvollen Quellen
Paper from responsible sources
FSC® C105338

If you have any concerns about our products,
you can contact us on
ProductSafety@springernature.com

In case Publisher is established outside the EU,
the EU authorized representative is:
Springer Nature Customer Service Center GmbH
Europaplatz 3, 69115 Heidelberg, Germany

Printed by Libri Plureos GmbH
in Hamburg, Germany